AF879557

THE AI TOOLKIT FOR TEACHERS MADE EASY:

10 SIMPLE STRATEGIES

TO SAVE TIME, PERSONALIZE LEARNING,
Empower Every Student to SUCCEED

-Even If You're Not Tech-Savvy

WASHINGTON PUBLISHING™ ENTERPRISES
Jacksonville • Silver Spring • Dominican Republic

Thank You!

FOR PURCHASING MY LABOR-OF--LOVE BOOK

THE AI TOOLKIT FOR TEACHERS MADE EASY

10 SIMPLE STRATEGIES TO
SAVE TIME, PERSONALIZE LEARNING,
EMPOWER EVERY STUDENT TO SUCCEED
-Even If You're Not Tech-Savvy

Thank you for your continuous support.
MANY BLESSINGS,

"Aunt Hattie"

(DR. HATTIE N. WASHINGTON)

Published by
WASHINGTON PUBLISHING ENTERPRISES
ahpkids@aol.com | www.aitoolkit4teachers.ai | 443-804-6545

PRAISE FOR

Dr. Washington's

The AI Toolkit For Teachers Made Easy

Although too often underrated, the profound importance and impact of our world's educators cannot be overstated. Among the most faithful and fruitful in this field of labor, Dr. Hattie N. Washington is endowed by our Creator with an insatiable appetite for learning and an impressive aptitude for leading.

Once again, she has expanded her toolbox and extended it to the world's educators, whether academic, religious, parental, corporate, organizational, or otherwise.

The emergence of Artificial Intelligence in our world necessitates professionals from practically every field familiarize themselves with its relevant applications and responsible integration. Dr. Washington, in her Toolkit, has provided an invaluable on-ramp, enabling AI novices to safely enter and successfully navigate this relatively new dimension of the information highway.

Be sure to gift copies to your friends and colleagues today, thereby expanding your success and enhancing the quality of life for those you serve.

Dr. Haywood A. Robinson, III
Pastor, The People's Community
Baptist Church in Silver Spring, MD

THE ENDORSEMENTS

"As a history and performing arts teacher, navigating the challenges and opportunities of integrating AI into the classroom, I am delighted to endorse Dr. Hattie N. Washington's The AI Toolkit for Teachers Made Easy. Her clear, practical strategies help streamline lesson planning and automate routine tasks, freeing up time to focus on engaging students and deepening their critical thinking. This toolkit empowers me to personalize learning, teach students how to ask meaningful questions with AI, and continuously improve my practice, making technology accessible and truly beneficial for all learners. Dr. Washington's work is an inspiring and essential resource for any educator ready to embrace the future of teaching confidently and purposefully.

JoAnna Ursal, MFA, MA
History & Performing Arts Educator,
Global Leadership Academy High School
(Title I), Jacksonville, FL

As a recent high school graduate, I'm genuinely thrilled to endorse my grandmother's book on using AI in the classroom. It's perfect for students like me who've grown up with tech and want learning to feel fresh and relevant. She demonstrates how AI can tailor lessons to you more effectively, spark your creative juices, and maintain honesty by fostering genuine engagement rather than taking shortcuts. Reflecting on my high school experience, where I put in a tremendous amount of effort to excel, her ideas would have made classes even more engaging, and I look forward to other students benefiting from her vision. What I love about her book is that it's both practical and inspiring, speaking directly to someone like me who's about to jump into college. My grandmother's passion for helping students and teachers shines through, showing how AI can boost curiosity and critical thinking. This book isn't just a guide; it's a bold look at how education can evolve, and I'm so proud to see my grandmother leading the charge for students like me.

Reagan S. Thomas
Granddaughters, Freshman,
TXST University, Austin, TX

"Dr. Hattie Washington's 'THE AI TOOLKIT FOR TEACHERS" is an indispensable guide for every K-12 educator eager to harness the transformative power of artificial intelligence. Recognizing that students are already engaging with AI, this book provides practical strategies to not only personalize learning experiences but also to empower teachers by reclaiming valuable time through automation. Dr. Washington expertly bridges the gap, offering a clear pathway for educators to confidently integrate AI tools, fostering a future-forward classroom where both teachers and students thrive in this exciting new era of education."

Dr. Jeffrey E. Berger, CEO,
Abundance in Optimum Wellness
Global Ambassador for DaMeta1

THE ENDORSEMENTS

"As someone who has utilized advanced technology during my 25-plus-year career in federal law enforcement, I can confidently say that the tools I once considered cutting-edge pale in comparison to the power and potential of today's rapidly evolving AI landscape. Dr. Hattie Washington's *THE AI TOOLKIT FOR TEACHERS MADE EASY,* is a timely and essential resource that equips educators—not only to understand this emerging technology—but to empower the next generation of leaders and thinkers to thrive in a world driven by innovation. This book is more than a guide; it's a blueprint for preparing young minds to navigate, adapt, and lead in the fast-paced reality of modern life."

Rev. Sean Thomas (Son-In-Law)
Federal Agent, US Dept. of Labor (Retired)

I wholeheartedly endorse Dr. Washington's "AI Toolkit for Teachers" as an essential resource for educators and students of all ages. Throughout her life, she has inspired others to leverage technology for positive change. This book empowers everyone—not just educators—to use AI tools to enhance their lives. The AI chapters equip readers to learn effectively and adapt to our rapidly changing world. This groundbreaking book simplifies tasks and improves productivity for all. Don't miss the chance to embrace the future with Dr. Washington's invaluable insights!

Willer Leveille
Director, Executive Internal Audit (Retired)
Banking Operations, Sogebank Group

Dr. (Aunt) Hattie N. Washington is a breath of fresh air for those impacting the marketplace, the seven mountains of influence, and more. This AI Toolkit for Teachers is a one-stop shop. As a global podcast host, it is efficient in helping entrepreneurs stay up to date with the latest technological shifts, enabling them to embrace their calling, assignment, and gift in an ever-growing world. I highly recommend this book for teachers, global entrepreneurs, single parents striving to support their children, and others. It is a toolkit that you deserve and need in everyday life, as you face the fundamentals of technology and want the opportunity to scale to the next level with ease.

Dr. Frances Ann Bailey,
International Podcast Host; 2025
National US Woman of Achievement

THE AI TOOLKIT FOR TEACHERS MADE EASY:

THE AI TOOLKIT FOR TEACHERS MADE EASY:

10 SIMPLE STRATEGIES

TO SAVE TIME, PERSONALIZE LEARNING,
Empower Every Student to SUCCEED

-Even If You're Not Tech-Savvy

DR. HATTIE N. WASHINGTON

WASHINGTON PUBLISHING™ ENTERPRISES

WASHINGTON PUBLISHING. ENTERPRISES

THE AI TOOLKIT FOR TEACHERS MADE EASY: 10 Simple Strategies to Save Time, Personalize Learning, Empower Every Student to SUCCEED -*Even If You're Not Tech-Savvy*

Published by:
WASHINGTON PUBLISHING ENTERPRISES
9938 Deercreek Club Road, East
Jacksonville, Florida, 32256 USA

Author Contact & Event Booking:
Dr. Hattie N. Washington, President & CEO of Washington Publishing Enterprises, is available for speaking engagements, workshops, training sessions, consulting, book signings, book club appearances, etc.
Email: ahpkids@aol.com
Websites: www.aitoolkit4teachers.ai | www.driven2succeed.com
Phone: 443-804-6545

Bulk Orders & Permissions:
For bulk purchase discounts or permission requests, please email: hwashington0001@gmail.com

Design Credits:
Book Covers and AI TOOLKIT Merch Logo Designs: Fayshel Graphics
Interior Images and Illustration Designed using Freepics.com – AI Tools - "OWN It!" Magazine & Publishing LLC.
E-Book Formatting and Creative Design: Marissa Bloedoorn, President & CEO, OWN IT MAGAZINE & PUBLISHING LLC
Paperback & Hardback books reformatted by Mahamud Noman, using the AI TEACHERS e-book design.

Library of Congress Control Number

Library of Congress Cataloging-in-Publication Data:
Washington, Dr. Hattie N.

THE AI TOOLKIT FOR TEACHERS MADE EASY: 10 Simple Strategies to Save Time, Personalize Learning, and Empower Every Student to SUCCEED—Even If You're Not Tech-Savvy / Dr. Hattie N. Washington p. cm

E-Book:	ISBN: 13: 978-1-950707-16-4 | 10: 1-950707-16-4
Paperback:	ISBN: 13: 978-1-950707-17-1 | 10: 1-950707-17-1
Hardcopy:	ISBN: 13: 978-1-950707-27-0 | 10: 1-950707-27-0
Audiobook (Audible):	ISBN: 13: 978-1-950707-18-8 | 10: 1-950707-18-8

Printed in the United States of America
10 9 8 7 6 5 4 3 2 1

PUBLISHER'S NOTE:

Washington Publishing Enterprises has made every effort to ensure the accuracy, proper licensing, and accessibility of all content, including third-party visuals, hyperlinks, and resources included in this publication.

This e-book contains a combination of the following image types:

- Licensed stock images selected and provided by the formatter.
- AI-generated visuals created with OpenAI's DALL·E and Freepix
- Free-use images from Pixabay, Unsplash, and Pexels under their commercial-use licenses by the formatter.

All images have been obtained with the proper licenses or permissions and are utilized in full compliance with the usage policies of their respective platforms. All trademarks and logos remain the property of their respective owners.

Please note: This e-book includes hyperlinks. Depending on your reading device, some links may not be clickable. If a link does not work, try copying and pasting the URL into your browser.

Washington Publishing Enterprises does not endorse external websites, advertisements, or services that may appear via third-party links. Use of such resources is at your discretion.

AUTHOR'S NOTE:

As an educator, author, and lifelong learner, I have included tools and platforms in this book that are accessible to all readers. Many of the resources referenced are available for unrestricted use. However, some sites may display ads or encourage you to upgrade to a paid subscription version.

Please know that these upgrades are entirely optional. In most cases, the free version is more than sufficient for the strategies and skills outlined in this book.

Always read privacy statements, use discretion when navigating websites, and avoid sharing personal information on platforms you don't trust. I hope these resources empower you to embrace modern technology and thrive in your journey of educational excellence.

✅....Even If You're Not Tech-Savvy.

"He who knows,
But knows Not How to express,
What he knows,
Becomes less sure,
As time elapses,
and may even wonder,
IF he ever knew at first."

--Dr. Hattie N. Washington

(Reinforcing the need for the VAKT Learning Styles Techniques)

****

DEDICATION

This book is lovingly dedicated to all the TEACHERS, educators, teacher's aids, school secretaries, custodians, bus drivers, parents, caregivers, and all other champions of children who believe in the unlimited potential within every student and who offer that extra Smile and encouraging word intentionally.

To my two devoted daughters — Dr. Charrell W. Thomas, MD, and Cheryl Y. Washington, Esq., who have always inspired me with their brilliance, compassion, and integrity. You embody excellence and purpose in every way. Thanks for ALWAYS being my great cheerleaders.

To my first foster son, Wayne Saunders, a brilliant IT Specialist at Microsoft — your success reminds us that love, structure, and faith can truly transform lives. You make me PROUD every day, knowing how much progress you've made over the years, and continue to make,

To my precious granddaughters — Cameron Monroe Thomas and Reagan Sinclair Thomas — may your curiosity, kindness, and boldness continue to light the way for generations to come.

And to all the foster young men who've walked through the doors of Aunt Hattie's Place into my (our) heart during the 22 years of rearing deserving fine young men, who didn't want a **"Handout,"** they just needed a **"Helping hand!"** Just know I am honored and feel highly favored to be referred to, even years later, as your **"Aunt Hattie!"** You are loved, and you matter, not only to me but to your entire "Village of Supporters." Go forth and continue to be **DRIVEN2SUCCEED.**

In conclusion, we extend our gratitude to educators, particularly those who serve the most vulnerable populations. I hope that this book provides renewed optimism, actionable strategies, and practical tools to enhance the learning experience for every student.

Love & Blessings

Aunt Hattie

ACKNOWLEDGMENT

I wish to personally thank the following people for their contributions and knowledge, helping in creating this book:

Thank you to my two daughters, Charrell Washington-Thomas, MD and Cheryl Washington, Esq. whom I am immensely proud and feel incredibly blessed to have had the sheer joy of rearing these two extraordinary young ladies; my fantastic, adopted Son, Wayne Saunders, whom I go to, still, for his incredible computer skills; my *favorite* son-in-law, Sean Thomas; and my amazing granddaughters, Cameron and Reagan Thomas. Thank you all for reviewing and endorsing my book, as well as reviewing posters and other promotional materials on social media and providing encouraging feedback that helped me complete this timely toolkit for teachers (and for anyone who interacts with children in today's ever-changing world of technology).

Thank you to my foster young men over the years. You have inspired me more than you will ever know to write about the impact of having computer labs in all three of my boys' homes in this book.

To all my supporters who endorsed my book: Marissa Bloedoorn, Pastor Jacqueline ("LuLu") Brown, Peggy Morris, Reba Burriss-Barnes, Cheryl Y, Washington, Esq., Charrell W. Thomas, MD, Wayne Saunders, Sean Thomas, Reagan Thomas, Valerie Leveille', Kelly Toaston, Lamar Myers, Sabine Vicioso-David, Dr. Jeffrey Berger, Dr. Frances Ann Bailey, Sharon Parker, Reba Burriss-Barnes, Peggy Morris, Charlene Day, Laticia Smith, Jessie (Jack) Handley, JoAnn Ursal,, Shawn (Papi) Hay, Fayshel, Cameron Thomas (Prayers),

I appreciate your faith in me and your unyielding support.

Thank my pastor, Rev./Dr. Haywood Robinson, First Lady Renee, and the entire congregation of The People's Community Baptist Church of Silver Spring, MD (whom I still watch virtually), for their continuous support, prayers, and encouragement. Now, a shoutout to the Julington Baptist Church here in Jacksonville, FL—where Rev. N. Jerome McClain is the Senior Pastor, and my son-in-law (Sean Thomas) is a minister there.

I want to extend my special gratitude and appreciation to Marissa Bloedoorn ("Lady M") of "OWN IT" MAGAZINE & PUBLISHING, LLC for her personable and invaluable expertise throughout the entire publishing process—from designing flyers for social media, posting on social media, press release, to creatively formatting the book, and whose creativity, design, and technical expertise brought these pages to life. Your attention to detail, commitment to this mission, and networking with the "OWN IT" family for their support are greatly appreciated.

Thus, I am also profoundly grateful to everyone in the "OWN IT" family and my "Village of Supporters" who contributed to the realization of this book. Your letters of endorsement, purchase of launch day books, written Reviews, words of encouragement, and prayers contributed to our reaching Amazon's #1 best-seller status in four categories.

To Dr. Jacqueline "LuLu" Brown, your prayers, partnership, and positive energy have uplifted me repeatedly, helping to promote, publicize, and share your other varied gifts with this project, including graciously agreeing to write the *Foreword.*

To Sharon Parker, Founder of ROASA, Inc., thank you for being a faithful supporter, connector, and purpose-driven leader who consistently makes a difference in the community, both locally and globally. Her mission with AMI (American Mothers Internationally) is far-reaching and empowering to all.

To Dr. Frances Ann Bailey, Founder and Host of Purpose Is Greater International Podcast, your platform and encouragement have lifted and expanded the reach of my AI movement work. I look forward to collaborating further and broadening our base to reach more people globally.

To Valerie Leveille and Kelly Toaston, who, in addition to writing an endorsement, took time on July 4th to come on a "Celebration For Teachers" Zoom Show, to share their views of the AI TOOLKIT FOR TEACHERS Book—A job well done!

A special shout-out and appreciation to Lamar Myers of AI & I--your vision and strategic insight helped bridge innovation with education in powerful ways; who is creating my proprietary interactive App, allowing us to extend further and delve deeper into AI and its virtual applications to reach all students. His gift and AI talents are noteworthy. I appreciate his generosity and interest in my AI mission and movement.

A special thank you to Brittany, my Coach at Publishing.com, as well as other coaches and fellow authors, for their valuable advice, training, and resources. You made the entire process a movement-driven and mission-focused learning experience.

And to every reader, teacher, and lifelong learner, thank you for being part of this movement. Let's continue to raise the bar and expect excellence from everyone.

Lastly, but in no means least, My Thank You, God Prayer:

I thank God for His almighty power and His favor in giving me the necessary vision, inspiration, wisdom, and wit to create this toolkit for teachers (and other caring individuals) to work smarter, not harder.

You allow things to happen to us to test our faith and reliance on You. You make this book happen for a reason, a mission, and a movement. This prayer is intended to express my appreciation, recognition, and exhortation to His name. My testimony and His revelation will be shared later.

May God richly bless each person who reads this book and who endeavors to empower every learner to BELIEVE, ACHIEVE, AND SUCCEED; thereby, "giving back" to the community and making a difference in this ever-changing world. Amen

With Love and Blessings,

Aunt Hattie

(Dr. Hattie N Washington)

I HEAR, and I FORGET. *(Auditory)*
I SEE, and I REMEMBER. *(Visual)*

But When

I DO, and I UNDERSTAND. *(Kinesthetics & Tactual)*
--Confucius, Philosopher & Reformer

[The meaning of the poem—As it relates to the VAKT (Visual, Auditory, Kinesthetic, and Tactual) Learning Styles]

"Wisdom is the principal thing; therefore get wisdom: and with all thy getting get UNDERSTANDING."
— Proverbs 4:7

NOTE: Prioritize wisdom and judgment, not just technical skill, when using AI in teaching.

CONTENTS

FOREWORD

As a former corporate technology executive with over four decades of experience leading global innovation and a licensed minister with more than 25 years of service in faith-based and underserved communities, I have seen firsthand the challenges and disparities that persist in our education systems, particularly in under-resourced districts. I am also profoundly aware of school boards' responsibility to ensure all students are equipped for a future shaped by technology.

I strongly endorse The AI Toolkit for Teachers Made Easy by *Dr. Hattie N. Washington.* This book is not just another instructional guide but a lifeline for educators navigating an overwhelming digital shift. Dr. Washington provides 10 simple, scalable strategies that help teachers implement artificial intelligence tools to save time, personalize learning, and boost student achievement—even without a tech background.

What makes this resource indispensable for school boards is its focus on equity, accessibility, and inclusion. It bridges the digital divide by empowering teachers in all schools, especially those in marginalized or underfunded communities, to harness AI in ways that are practical, ethical, and aligned with student success outcomes.

This book delivers vision and action for school districts striving to future-proof their instructional models while supporting teacher wellness, closing achievement gaps, and honoring diverse learners.

In every role I've held—from corporate boardrooms to community classrooms—I've led with a mission to "innovate to add value," ensuring that every solution advance technology and uplifts people, especially those in underserved communities.

Dr. Washington does this indisputably, offering practical, heartfelt guidance that meets teachers where they are, whether tech-savvy or not. Her ability to "innovate to add value" in education, especially for communities often left behind, is timely and transformational. I am honored to endorse this groundbreaking book and encourage school boards to champion its adoption as part of their strategic investment in equitable, future-ready learning.

"Peace, Love, Light, Joy, Ease, & Flow"

Jacqueline Lulu Brown

Founder & CEO – Revolution Ascension LLC

PREFACE

A Heartfelt Welcome from Dr. Hattie N. Washington

Welcome, dear Teacher and Educator,

Whether you are an experienced educator, a novice teacher, a technology-oriented leader, or someone who needs assistance with classroom equipment, this book has been **written with you in mind** to address your needs. With deep love, admiration, and appreciation, I honor your commitment to a profession that is more than a job—it is a mission, a ministry, and a sacred trust.

As someone who has walked your path—as a teacher in both regular and special education, as a Child Development Specialist, Assistant Superintendent, University Professor, and most proudly, as a mentor and cheerleader to countless educators, I've seen firsthand the exhaustion in your eyes and the fire in your heart. I have also witnessed how the right tools, used wisely, can reignite your joy, restore your energy, and redeem your time.

This book is a lifeline. A guide with AI by your side. Your new companion for navigating the ever-changing landscape of education with confidence, creativity, and clarity.

Why This Book Matters Now More Than Ever

Education is at a tipping point. We're grappling with historic teacher shortages, widening learning gaps, increasing neurodiversity, and rising demands for accountability. At the same time, Artificial Intelligence is surging into our classrooms—exciting, intimidating, and full of potential.

Let's be clear: **AI is not here to replace you. It's here to partner with you.**

But the truth is, most teachers weren't trained in how to use AI effectively, and many who try end up feeling more overwhelmed than supported.

That's where this book comes in.

This is not a tech manual. This is a soul-centered, brain-smart, and heart-forward guide to:

- Saving your precious time
- Personalizing learning for every student
- Collaborating with AI to make smarter, more informed decisions
- Meeting the needs of SPED, Title I, and Gifted learners
- Integrating AI with SEL, equity, and genuine human connection

And we'll do it all without jargon, without pressure, and without assuming you're a tech expert—just thoughtful strategies, practical tools, and real support.

Let's do this together—with **faith, favor, focus**, and the **fierce determination** that *no student gets left behind.*

With love and purpose,

DR. HATTIE N. WASHINGTON

("Aunt Hattie")

"AI rubrics and dashboards aren't about catching students in a mistake; They're about building a bridge toward better practice & stronger outcomes."

INTRODUCTION

"You are Not a Teacher Until Students Have Learned What You Taught--By Any Means Necessary"

— Dr. Hattie N. Washington

A bell rings down the hallway, and in Room 207, a teacher stands in front of a whiteboard. There's a stack of ungraded papers leaning dangerously to one side on her desk. Her laptop hums with tabs open for lesson plans, emails from parents, and a spreadsheet tracking which students need extra support.

She's got a student with an IEP sitting in the front row, a new English learner quietly trying to keep up, and a gifted student in the back who finished the assignment before the directions were even out of her mouth. There's a clock ticking overhead and a mental checklist that just keeps growing.

Sound familiar? I've loved it, and I know you have too. The demands of today's teachers are real, and they're not getting any smaller. More paperwork. More students with unique needs. More pressure to personalize, differentiate, and connect. And then, just when you thought you were getting a handle on it all, everyone starts talking about "AI in the classroom." Like you needed one more thing on your plate, right?

Here's the promise of this book: I'm not here to add to your stress. I'm here to hand you a toolkit. Not in the abstract, pie-in-the-sky sense, but a real, practical set of strategies you can use tomorrow to save time, personalize learning, and reach every student—even if you still call your laptop "the computer thingy." Even if you're not tech-savvy. Especially if you're not tech-savvy.

Aunt Hattie

You might be wondering, "Who is this person, and why should I trust her with my precious planning time?" Let me introduce myself. I'm Dr. Hattie N. Washington—an educator, tech enthusiast, and, yes, a proud mom and **"Aunt Hattie"** to scores of foster boys.

More than thirty years ago, I hosted a TV show called Computers for Kids, teaching families across the DC, Maryland, and Virginia area that technology wasn't something to fear. I set up after-school computer labs in churches and installed computers in the homes of my three foster boys. I guided low-achieving students with special education, Title I, ADHD, and others toward a brighter future. In fact, due to early exposure to computers, my first foster son went from having no computer experience to majoring in Information Technology in college and now has landed a job at Microsoft.

I've spent my life helping teachers, students, and parents bridge the gap between technology fear and greater possibility.

This book serves as your roadmap.

My goal? To help you become an AI-ready educator. You'll find ten simple, actionable strategies, each one a step toward working smarter, not harder.

Four Phases of Real Teaching and Learning

The book is organized around the **four phases** of real teaching and learning: **PLAN, TEACH, ASSESS, and REFLECT.**

No theory without action. No jargon without translation. Just tools you can grab and use right away.

Whether you're a teacher, administrator, parent, or student, this book is for you. Yes, you.

No matter what your background, your grade level, or your comfort with technology. I've designed every chapter to be accessible, practical, and—dare I say—fun.

Special education, Title 1, ADHD, SEL, ELL, gifted, or just plain overwhelmed? You'll find something here that resonates with your needs and those of your students.

Each chapter stands alone.

You can jump to the part that matters most right now. Need help with lesson planning? There's a strategy (and templates) for that. You want to personalize learning without cloning yourself? It's covered. Dreaming of automating grading so you can reclaim your evenings? Yes, please.

Here's a quick peek at the ten strategies we'll explore together:

- Simplify lesson planning with AI tools.
- Personalize learning paths for every student.
- Automate grading and feedback to save your sanity.
- Differentiate instruction, even when your class feels like a jigsaw puzzle.
- Engage students creatively—bring some spark back to learning.
- Strengthen critical thinking, not just rote answers.
- Integrate social-emotional learning and mental wellness.
- Collaborate with classrooms (and experts) around the globe.
- Utilize AI analytics to evaluate growth and adjust instruction in real-time.
- Refine your teaching with continuous quality improvement, powered by fundamental data.

Every chapter comes with sample lesson plans, VAKT (Visual, Auditory, Kinesthetic, Tactile) templates, and real classroom scenarios. Wondering how this works for a student with ADHD? Or an ELL learner? I've included those stories, too. You'll see how each strategy plays out in the real world.

Now, let's talk about the elephants in the room: "But I'm not tech-savvy." "Our school has no budget for fancy gadgets." "Is AI going to take my job or make things harder?" You're not alone in asking. I've heard it from teachers who still keep chalk in their desks for emergencies, to parents who think AI is just robots taking over the world. The truth? You don't need to know Python or have a robot assistant.

These AI tools are free (or low-cost). And, when used appropriately, AI lets you do what you do best—build relationships, inspire curiosity, and help students grow.

The structure of this book mirrors what real teaching looks like—dynamic, cyclical, and always in motion.

There's No One Right Way To Use This Toolkit.

You can start with Planning, jump to Assessment, or dip into Reflection when you're ready. There's no one right way to use this toolkit. **You're in charge.**

To make things even easier, I've packed this book with ready-to-use checklists, study guides, IEP templates, and a glossary to cut through the tech-speak. No more Googling acronyms during your lunch break. And in the back? You'll find a set of AI brain teasers (because learning should be fun) and a glossary to keep handy for quick reference.

My Invitation To You

Here's my invitation to you: join me on this journey. Let's move from being overwhelmed to being empowered, from hesitant to confident. You don't have to do it all at once. You just have to take the next step. Together, we can help every student, regardless of their background, needs, or dreams, believe, achieve, and succeed.

Ready? Let's get started.

OVERVIEW

This is not a tech manual—

IT'S AN EMPOWERMENT GUIDE FOR TEACHERS IN EVERY SETTING.

Inside, you'll find:

- ✅ 10 easy-to-use AI strategies
- ✅ Step-by-step instructions with screenshots
- ✅ Real-life classroom examples
- ✅ Adaptations for Special Ed, ELL, and advanced learners
- ✅ Brain teasers, checklists, and hands-on activities
- ✅ VAKT learning style supports
- ✅ Bloom's Taxonomy-aligned objectives

WHETHER YOU'RE A TECH NOVICE OR ALREADY EXPLORING AI, THIS BOOK MEETS YOU EXACTLY WHERE YOU ARE.

OVERVIEW

How to Use This Book: A Quick Guide for Busy Educators

Here's how to get the most from *THE AI TOOLKIT FOR TEACHERS MADE EASY:*

1. **Start where you are. Each chapter builds on the last, but you can jump anywhere.** Overwhelmed? Begin with Chapter 5 (AI as Your Time-Saving Co-Teacher). Curious about creativity? Jump to Chapter 6.
2. **Use the Benchmarks.** At the start and end of every chapter, you'll find benchmarks for Awareness, Implementation, Mastery, and CQI (Continuous Quality Improvement). **These help you self-assess your skills and growth over time.**
3. **Try It, Don't Just Read It.** Each chapter includes real AI prompts, templates, sample lesson plans, graphic organizers, printable tools, and even links to live demos and downloadable resources.
4. Think Differentiation. Every Strategy is Adaptable For:
    - General Education
    - Special Education (SPED/IEPs)
    - Title 1 & Under-Resourced Schools
    - Multilingual Learners (ELL/ESOL)
    - ADHD/SEL
    - Gifted & Talented (GT)
5. **Have fun with the Study Guide & Workbook!** At the end of each chapter, you'll find brain teasers, puzzles, acrostics, reflection prompts, and games for both students and teachers. Learning should be joyful.
6. **Come back often. AI is evolving.** So are you. This book is not a "one and done" read. Mark it up. Bookmark pages. Share it with colleagues.
7. Grow with it.

Demystifying AI For Every Educator

Chapter One

CHAPTER 1:

DEMYSTIFYING AI FOR EVERY EDUCATOR

ACTIONABLE, MEASURABLE OBJECTIVES FOR CHAPTER 1

- ✓ **OBJECTIVE:** Educators will describe and explain at least three practical classroom applications of AI using visual aids, peer discussion, and interactive simulations.
- ✓ **VAKT ALIGNMENT:** Visual: Diagrams and infographics on AI terms; Auditory: Peer discussions and video tutorials; Kinesthetic: AI simulation walkthroughs; Tactual: Printed guides and charts.
- ✓ **BLOOM'S LEVEL:** Knowledge, Comprehension
- ✓ **INCLUSIVE SUPPORTS:** Supports all learners through multi-modal delivery and scaffolded AI literacy content.

Back In My Early Days: Dinosaur Computer

Let's take a quick trip back to my first year of teaching special education students. I had one computer in my classroom that had a floppy disk (I am dating myself, but I want you to get the point how far technology has come)

That one computer was primarily used to reinforce math and language arts skills and to type the updated Individualized Education Plans (IEPs) of my students.

I remember feeling intimidated by that old dinosaur machine then, trying to learn the literacy of the dawn computer, all while attempting to decipher what to write on the IEP to ensure each of my students would accomplish the objectives on their plan, all while including their observed learning style to their related services, called accommodations. Because the process of entering data for each of my 12 students was time-consuming and took time away from my direct instruction to the students, it didn't allow them to spend time on the computers for their subject learning and reinforcement.

I completed some IEP documents and lesson plans after school on that computer. However, the majority were taken home to allow for adequate time to consider the best decision for each student, as well as the overall class. Although I enjoyed teaching and held my students in high regard, I aspired to effectively reach each student according to their individual needs and circumstances. Nevertheless, I often felt overwhelmed by the responsibility of the required paperwork and the personalized plans.

Thus, I can only imagine how teachers feel today, with the demand to do more with less and to reach all students, and to keep up with all the federal, state, and local requirements of special education students (who are now mainstreamed in the regular classroom.

Background History

With the passing of the IDEA (Individuals with Disabilities Education Act), first signed into law in 1975 as the Education for All Handicapped Children Act (EHA), mandating that all students with disabilities (SWDs) receive free and appropriate public education (FAPE) in the least restrictive environment (LRE) possible. A student's FAPE and LRE are established through a process that produces an IEP for each student.

I took a moment to provide you with that background information on special education and the federal, state, and local requirements, so that you will know that I empathize with the requirements and the voluminous paperwork that regular education teachers now must complete while teaching and reaching every student.

This toolkit should lighten your load by providing a selection of free AI tools, software, links, and resources to assist you with the various tasks and responsibilities that teachers face on a daily basis. As I indicated earlier in the book, each chapter will contain some of these tools, as well as more in the directory in the back of the book:

Reminder: (See the Disclaimer in front of my book for all free AI Tools & software)

AI Tools & Resources for Chapter 1:

AI Free Tools Aligned with The Four Instructional Phases **(PLAN; TEACH; ASSESS; & REFLECT)**, the Chapter Content, and the Indicated Strategy

(Disclaimer Note: Skip Ads that pop up in most free links, apps, and software. Although many free links offer advanced plans for a monthly fee, there are sufficient free links available for your students. Try other links under the relevant topic. If any of the links are not active, try cutting and pasting the link into your browser.)

Phase I: PLAN

Chapter 1. Demystifying AI for Every Educator

- MagicSchool.ai
    - **Description:** A free AI-powered platform with 60+ tools for lesson plans, rubrics, IEPs, parent communication, and more—designed specifically for educators.
    - **Link:** https://www.magicschool.ai/
- EdoAid
    - **Description:** Creates quick, AI-generated learning plans and classroom activities tailored to student outcomes and Bloom's Taxonomy.
    - **Link:** https://edoaid.com/
- Educare
    - **Description:** Educator-friendly AI tool for generating scaffolded lessons and differentiated resources aligned with multiple standards.
    - **Link:** https://educare.ai/

OTHER AI RESOURCES & SOFTWARE REINFORCING THIS CHAPTER:

- AI for Education by ISTE *(see the details about ISTE in the Glossary):*
    - **Description:** Intro course helping educators understand AI in simple terms.
    - **Link:** https://iste.org/AI (*see more info about ISTE in the Glossary)*

- Google's AI for Education
  - Description: Free AI tools and courses from Google tailored for teachers.
  - Link: https://edu.google.com/
- Microsoft Learn AI for Educators
  - Description: Microsoft's free educator training for AI integration in classrooms.
  - Link: https://learn.microsoft.com/en-us/training/educator-center/

So, let's start this chapter with the truth: you do not have to be a computer whiz to understand AI--Or to use it for your students' benefit.

What AI Means in Your Classroom—No Jargon, Just Practice

AI can sound intimidating, but all it means for you is this: intelligent computer helpers working behind the scenes. Think of AI as a digital teaching assistant who doesn't drink your coffee or loses your favorite pen. Instead, your new helper sorts stacks of paperwork, fetches ideas when your brain goes blank, and reminds you that there are only so many hours in the day.

AI isn't here to take your place at the front of the room. It's here to handle those repetitive tasks that eat up your time (and maybe some of your patience). If you've ever wanted to clone yourself during grading season or wished for a magical reading-passage fairy, you're in the right place.

Let's skip the technical mumbo-jumbo and see what this looks like in real classrooms. Say you've given a quiz to three classes. Instead of spending your evening hunched over a stack of papers and red pens, platforms like Quizizz or Edulastic can grade those quizzes in seconds.

These tools use AI to recognize answers and tally scores, freeing you up to plan tomorrow's lesson—or, more importantly, catch up on some much-needed rest. Suppose you need to create reading passages at different levels for your diverse learners.

In that case, you can use an AI tool like ChatGPT or MagicSchool to whip up three versions of the same story: one for your emerging readers, one for your on-level students, and one for that gifted kid who breezed through last week's assignment before you finished explaining it:

No more one-size-fits-all handouts or last-minute scrambling for materials.

Now, let's talk about something every teacher dreads: scheduling parent-teacher conferences. You send out forms, parents email back with eleven different date requests, and suddenly, you're using sticky notes as a scheduling system. Enter AI-powered calendars that can automatically suggest meeting times based on everyone's availability, send reminders, and even reschedule when life happens. Suddenly, your desk is less cluttered, and your stress level drops a notch.

But here's the critical part—AI does not replace what only humans can do. Your new digital assistant can grade quizzes, but it can't sense when a student needs a pep talk after a tough day or spot the quiet look of pride when a struggling learner finally "gets it." It doesn't know that sometimes kids fib about "the computer ate my homework." That's all you. AI can suggest ideas, but it doesn't have your creativity or your understanding of what motivates your students. You bring empathy, humor, and ethical judgment to every decision; a computer simply follows instructions.

Chances are, you're already using some form of AI in your classroom, even if you don't call it that. When you use spell check in Google Docs or Microsoft Word, that's AI quietly fixing typos before anyone sees them. If you ask Siri or Google Assistant to set a timer for group work or remind you about library day, yep—that's AI lending a hand. These familiar tools make newer AI applications feel less intimidating and more like a natural next step.

Glossary Sidebar

- AI (Artificial Intelligence): Computer programs that can "think" or solve problems in ways that mimic humans.
- Digital Assistant: A computer program designed to help you manage tasks.
- Automated Grading: Software that checks student work and assigns scores.
- Differentiation: Adjusting instruction to meet diverse learning needs.
- Natural Language Processing: How computers understand and generate human language (like when ChatGPT writes text).

YOU Remain the Heart of the Classroom

In short, AI is here as the helpful assistant who never calls out sick or leaves dirty coffee mugs behind. You remain the heart of the classroom, the connector, motivator, and decision-maker. With these innovative tools at your side, you can focus on what drew you to education in the first place: helping young people grow and thrive.

Overcoming "I'm Not Tech-Savvy": Real Teacher Stories & Simple Mindset Shifts

A few years back, I sat across from a veteran special education teacher—let's call her Ms. Howard—who had been in the classroom longer than some of her colleagues had been alive. She was the steady hand during fire drills, the queen of visual schedules, the one who could spot a meltdown brewing from down the hall. Technology, however, had always made her palms sweat. When my district rolled out a new tool for IEP progress monitoring, she confessed her anxiety in the teacher's lounge, whispering, *"I can barely keep up with email—how am I supposed to do this?"*

"Plus," she said, *"I have tenure and will be retiring in a few years, and they can't make me."*

After offering words of encouragement and demonstrating the process, using one of her students as a real-life example, she decided to give it a try, opting to use the tool for just one student's progress report instead of all of them.

She watched a three-minute video tutorial, poked around the dashboard, and let herself fumble through the first run. The following week, she showed up—less frazzled, more curious—and told me, *"It took me fifteen minutes instead of my usual hour. I didn't break anything. And my report looked sharp!"* Her story didn't end with her becoming some overnight tech superstar; she still writes notes on sticky pads and keeps backup paper copies for everything. But she proved to herself that it's possible to learn something new, even when technology feels intimidating.

You don't have to leap into the deep end. Consider trying AI for a low-stakes task, such as attendance or morning announcements. Make it a goal not to "get it perfect" but simply to experiment.

✓ DIP YOUR TOE IN FIRST, AND YOU WILL SEE IT'S NOT AS DIFFICULT AS YOU THOUGHT.

Here's a quick checklist for your first dip:

FIRST STEPS CHECKLIST:

- ✓ Choose one easy task (such as attendance, reminders, or a simple form).
- ✓ Try the tool just once this week—no pressure if it doesn't work out.
- ✓ Click on a free AI tool, watch a brief how-to video, or ask a fellow teacher for assistance with a demo.
- ✓ Celebrate any progress (even if you only manage to log in).

MEET MRS. SANCHEZ, AN ESL TEACHER.

Now, meet Mrs. Sanchez, an ESL teacher, whose classroom is brimming with students from a dozen different countries. Vocabulary practice had always been an uphill climb for her—one worksheet never fit all of her students, and Google Translate sometimes produced more giggles than grammar.

FROM CONTROL TO CURIOSITY

She came across an AI translation extension that allowed students to practice words in their native language before switching to English. At first, she worried about losing control or confusing the lesson plan, but she watched her students light up when they could check their work and play with language independently.

She admitted later, *"I stopped being the language gatekeeper and became their learning coach instead."* Her willingness to shift from control to curiosity allowed her students to blossom—and they began teaching each other new words in languages she didn't even speak.

Mindset: Progress Matters More Than Perfection.

This is where mindset becomes everything. The real trick isn't about learning tech; it's about allowing yourself to be a beginner again. Progress matters more than perfection.

Failing Forward—A Detour Rather Than A Roadblock

When you stumble—because you will—think of it as "failing forward--A detour rather than a roadblock." Every misstep is just another way to learn. This simple shift frees you from paralyzing pressure to be perfect on day one. Your students don't expect you to know everything, anyway; in fact, they often enjoy helping you figure things out.

"Tech Buddy-Snack & Click" Sessions

If you're still feeling especially hesitant, which I thoroughly understand, I suggest choosing and bringing a colleague, a "tech buddy," along for the ride. "Tech buddy" sessions after school can turn learning into laughter instead of dread.

You can start "Snack & Click" sessions where teachers and their "tech buddy" can bring snacks and explore new AI apps together—no formal agenda, just mutual support (and sometimes cookies). The bravest among them could project their screens and try something risky while the rest could cheer them on or offer rescue when the snacks were all finished.

Teachers sometimes forget that modeling learning is just as important as modeling content knowledge. I used to make that mistake myself, especially in my early years of teaching, when I was also teaching special education students. I was busy teaching the WHAT, not the HOW. I found student readiness (physically, emotionally, and psychologically) was just as crucial to receiving the great lesson I thought I had planned out meticulously, even though all of the students were on various readiness levels with different learning styles. Technology saved the day (although it was not yet called AI, but technology nonetheless).

From Perfectionism to Experimentation

When you admit you're trying something new with AI and your students see your process, your struggle, your questions, even your funny mistakes, you allow them to take risks too. "Let's see if this tool works," you might say. Or even better: "Who can show me where this button is?" Suddenly, your classroom culture shifts from one of perfectionism to one of experimentation.

Believe it or not, many of your students are already quite savvy with various aspects of AI—though they may not refer to them as AI—such as TikTok, Instagram, Microsoft editing, Grammarly, and other forms of high tech. They use these tools daily seamlessly and think nothing of it. They were born in a high-tech world. Therefore, let them take the lead in understanding some of the AI tools you are trying to learn, and simultaneously use them to teach with.

It's Intuitive, Mom

I unapologetically used to call on my genius foster son, Wayne, to help me figure out various glitches and issues with my computer. It seemed effortless for him. He would call me and put my computer or laptop on something called "Team View," and then take control of my cursor, moving it around and fixing whatever the problem was. Incredible! I asked him one day, *"How did you get so smart with computers and know exactly what to do by troubleshooting?"* to which he matter-of-factly responded, "It's intuitive, Mom."

Due to my early exposure to computers, I ended up majoring in Information Technology (IT) in college and now work for Microsoft.

A quick mental exercise for those feeling left behind: recall how many things you do now without thinking—navigating smartboards, using electronic gradebooks, sending mass parent emails—that once seemed impossible. AI is just another step on that same path. If you ever feel stuck or embarrassed about not knowing something, remember Ms. Howard and Mrs. Sanchez; their courage wasn't in knowing everything but in being willing to learn right alongside their students.

Try picking one small thing this week, something that feels manageable, and let yourself be curious rather than critical. The more you play, the less scary it becomes. And if you find yourself laughing at your own mistakes? Even better. That's how real growth begins—one imperfect click at a time.

Debunking AI Myths—Separating Fear from Fact in Education

AI myths in schools spread quickly, especially after a tech mishap, with the most common being, *"AI will take my job."* This anxiety often surfaces in staff lounges, but it calls for a closer examination. AI in education isn't about replacing essential human connections; it's about automating repetitive tasks, such as sorting assignments or scheduling, while also allowing educators to notice when a student looks upset.

Even the best classroom tech can't walk a student to the nurse or calm someone nervous about a test. The real story: AI gives you time back, it doesn't replace you.

Myth: "AI is Only for STEM Teachers."

A persistent myth says, *"AI is only for STEM teachers."* While science and math often get attention, AI is subject-agnostic. Meaning, whether you teach music, history, world languages, or art, AI can support your work.

World language teachers use AI translators to overcome language barriers, while art teachers encourage creativity through AI-generated suggestions.

An English teacher can utilize AI to analyze student writing and provide personalized feedback that students can easily read. AI adjusts to fit your subject and needs—it's not limited by the topic you teach.

AI Is Too Expensive For My School

The "AI is too expensive for my school" myth is also widespread. Having worked in budget-tight Title I schools, as well as serving as the assistant superintendent of an all-city-wide Title I district, I now know that many practical AI tools are either free or offer solid trial versions.

From translation plug-ins to auto-grading, it's possible to start small at no cost. One administrator from a small rural district said, "We began with three free tools and some YouTube tutorials—it was about experimentation, not budget." The claim that only wealthy schools can access AI is simply outdated. A smartphone and internet connection are often all you need to get started.

A common concern for parents and teachers is, "AI is unsafe or will make mistakes that harm students." Any tech can glitch, but responsible AI use means having checks in place and not letting apps run unsupervised. Schools should adhere to privacy rules and ethical standards (see Edutopia's guide, "AI and the Law: What Educators Need to Know"). Good educators constantly review AI's output, especially when it involves grades or student records. If you are concerned about bias or errors, professional oversight is key. Computers may calculate, but you know when a student needs another chance because of unique circumstances.

Some numbers: In a district pilot, teachers using free AI lesson generators saved about four hours a week on prep. Another school saw a 35% rise in parent engagement with AI-powered communication tools that automated reminders and translations. More than 90% of administrators surveyed felt more willing to try new tech after seeing how user-friendly these tools had become, regardless of cost or their initial tech concerns.

What If? Box: Addressing Skeptic Questions

Q: What if AI makesa gradingmistake?

A: Like a calculator, AI can slip if the data is poor or unclear. Always double-check critical results; treat AI as a draft checker, not the final decision-maker.

Q: What if my schoolcan'taffordexpensiveAI programs?

A: Try free options—many platforms offer no-cost plans or educator discounts. Check out MagicSchool and Google for Education for budget-friendly tools.

Q: What if parentsworry aboutprivacy?

A: Be transparent about your tools. Share your privacy policies, explain data safety, and invite family members to ask questions early.

Q: What if my subject isn't "techy"?

A: AI is for all subjects—from generating reading questions in English to helping with differentiated PE instruction. It's about saving time for your core work.

The real advantage of AI lies in using these tools as partners, amplifying strengths, identifying blind spots, and freeing teachers to do what only humans can: care deeply and teach with heart.

Assessing Your AI Readiness: Self-Checklists And Growth Roadmaps

Standing at the edge of something new, even if it's not your first rodeo, can make your palms sweat. That's perfectly normal. Before you reach for any shiny new tech, it helps to know where you stand right now, what feels like home, what feels like outer space. I've seen teachers of every age and background ask themselves, "Am I ready to add AI to my teaching toolkit?" There's no correct answer, only an honest one. This is where a self-check comes in handy.

Take a moment, grab a sticky note or the margin of your plan book, and jot down your answers as you read through these prompts. How often do you use technology in your lessons? Do you feel curious or cautious when someone mentions a new app? Are you the first to try a digital tool, or do you wait for that brave soul down the hall to test it first? Do you make time for professional learning—even if it's just a quick YouTube video or tip from a colleague?

Now, let's put this into practice. Here's a quick self-checklist:

- ✓ I can confidently use email and digital gradebooks.
- ✓ I've tried at least one new tech tool in the past year.
- ✓ I look for ways to make repetitive tasks more manageable.
- ✓ I'm open to learning from students, colleagues, or family.
- ✓ I have joined (or would consider joining) an online teacher group.
- ✓ I set aside time, even ten minutes a week, to explore something new related to teaching.

If you nodded along with none or just one, you're in the **"Getting Started"** camp—nothing wrong with that. Two or three checkmarks? You're **"Building Confidence."** If you breezed through most of them, you're **"Ready for More."** Each stage deserves its next steps—no skipping ahead required.

For those just starting, begin with the basics. Choose one task that feels safe—maybe using an AI-powered calendar for reminders or exploring a single app recommended by a trusted friend. Keep expectations low; this is about building comfort, not racing to the finish line. Beginner-friendly tutorials are everywhere, and sometimes the best learning happens when you click around without fear of **"breaking"** anything.

Remember, many tools are designed to be intuitive for everyone; even your grandmother could give them a whirl.

If you're building confidence, stretch a little further. Try using AI to generate quiz questions or analyze student responses. Explore peer support groups—many teachers have found Facebook groups or ISTE forums (International Society for Technology in Education) helpful for swapping ideas without judgment. Sometimes your best advice comes from someone across the country who's already weathered those first awkward moments.

For those ready for more, think about deepening your knowledge. Advanced users may want to explore certification programs or specialized webinars on integrating AI more fully into lesson planning and student assessment. Offer to mentor someone else starting; teaching others cements what you know and builds a community spirit in your school that leaves no one behind.

Support doesn't stop at self-reflection. Seek out networks—whether online cohorts or a district **"AI Champion"** who acts as a resident expert. Small schools may rotate this role, while larger districts often have teams dedicated to digital learning. Don't underestimate informal circles either; sometimes the best advice comes over coffee in the break room or during an after-school chat.

I want to pause and say: your growth with AI (or anything else) will never be a straight line. Some weeks you'll feel like you're flying; others, it may feel like you're dragging your feet through digital cement. This isn't about arriving at some imaginary finish line where you know everything there is to know about AI; it's about building comfort and confidence over time. The landscape of education is constantly shifting, so allow yourself to check in regularly with yourself. What felt scary last month might feel like second nature next semester.

Here's a reflection exercise for when you hit those moments of doubt: Recall an area where you once felt shaky but now feel sturdy—whether it was managing group work, learning a new curriculum, or just surviving your first year. Growth sneaks up on us while we're busy living our daily routines.

Keep your self-assessment handy as a living document. Revisit it when new tools appear on your radar or when school priorities change. Celebrate small wins and acknowledge setbacks as part of the process. You aren't expected to know it all overnight; your willingness to adapt already puts you ahead of the curve.

Support is all around if you know where to look—no one has to walk this path alone. Teaching is challenging enough without feeling isolated in your learning. By tuning into your readiness and seeking out community resources, you turn uncertainty into possibility and move one step closer to feeling at home with AI, no matter where you started today.

Celebrate Small Wins & Acknowledge Setbacks As Part Of The Process.

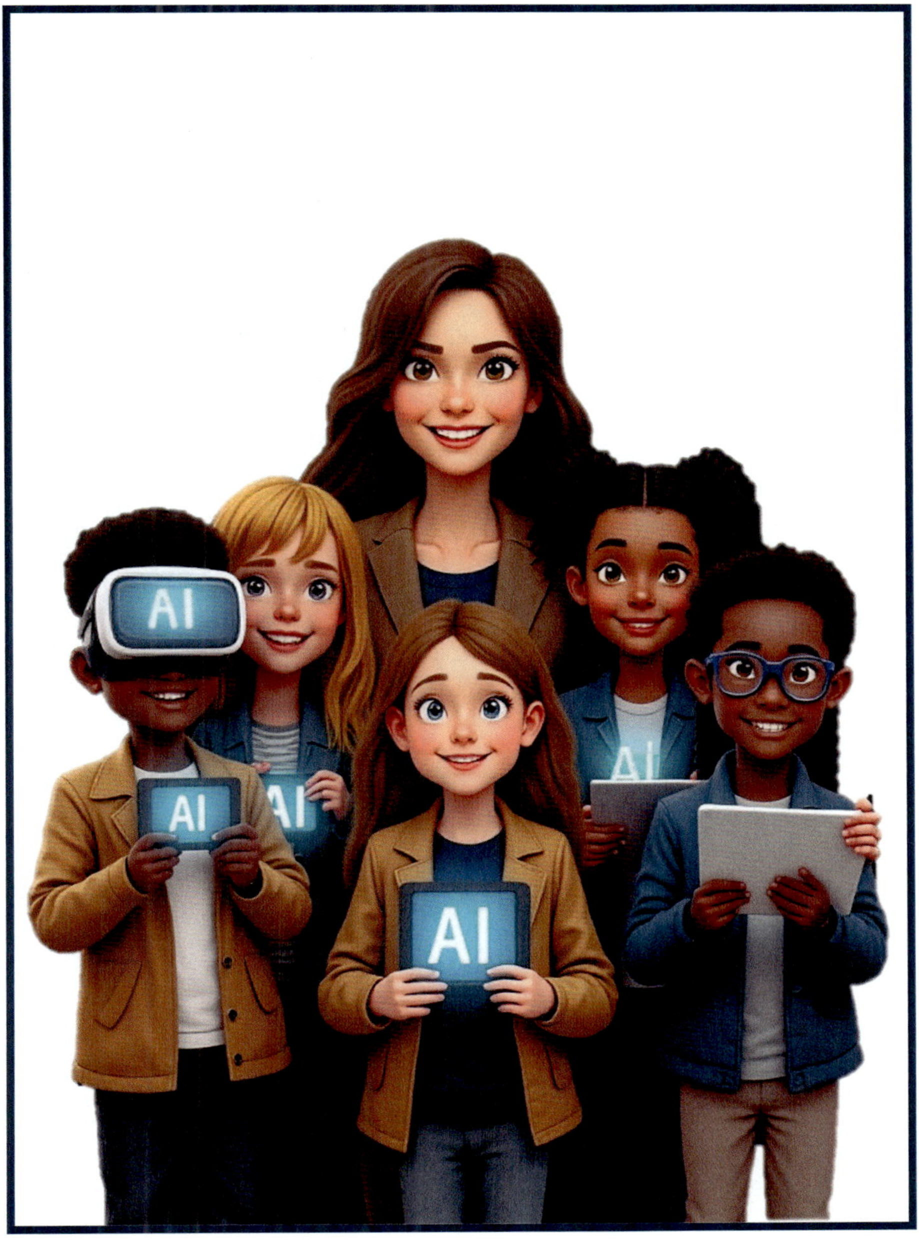

BUILDING A SAFE & ETHICAL AI FOUNDATION

CHAPTER TWO

"The rise of AI should mean the rise of human purpose."
— Demis Hassabis, CEO of DeepMind

CHAPTER 2:

Building a Safe & Ethical AI Foundation

"When technology integration is at its best, a child or a teacher doesn't stop to think that they are using a tech tool—it's just part of the learning process."
— National Education Technology Plan.

- ✓ **OBJECTIVE:** STUDENTS WILL EVALUATE THE ETHICAL IMPLICATIONS OF AT LEAST TWO AI SCENARIOS USING DIGITAL STORYTELLING OR GROUP ANALYSIS TOOLS.
- ✓ **VAKT Alignment:**
- ✓ **VISUAL:** Digital stories or case-study animations;
- ✓ **AUDITORY:** Recorded ethical debates;
- ✓ **KINESTHETIC:** Role-play activities;
- ✓ **TACTUAL:** Printable scenario cards.
- ✓ **BLOOM'S LEVEL:** Analysis, Evaluation
- ✓ **INCLUSIVE SUPPORTS:** Provides role-play and visual content for diverse learners, including ELL, SPED, and Gifted students.

Student Data Privacy—Concrete Steps to Keep Information Safe

Imagine a typical classroom day: you're introducing a new reading app or an AI-powered math platform. Excited students log in, but behind each username lies a trove of personal information.

Every click, response, or chat forms part of their digital footprint—details including names, birthdays, photos, emails, scores, and sometimes even family data. This student's data reflects both learning progress and personal identity.

AI tools often require a substantial amount of this data to function, such as analyzing progress, recommending steps, or sorting assignments. With these benefits come real responsibilities.

STUDENT DATA GENERALLY FALLS INTO THREE MAIN CATEGORIES:

- Personal identifiers: names, birthdays, student IDs, contact information.
- Academic records: grades, test scores, attendance, teacher feedback
- Behavioral analytics: participation patterns, app usage, time spent on questions.

Understanding these types helps educators know what needs protection. AI tools might use any or all of this information, making privacy not just a tech concern, but a critical safety and trust issue.

To protect student data when using AI tools, start with some basic protocols. First, check for compliance with FERPA (Family Educational Rights and Privacy Act) and COPPA (Children's Online Privacy Protection Act) for students under 13. These foundational laws show if student data receives basic protection (see APA list: 1). Next, review the tool's privacy policy. Look for definite statements that the data won't be sold or shared for advertising purposes if you encounter vague phrases like "improving services" or "sharing with partners." Exercise caution.

USE THIS QUICK CHECKLIST WHEN EVALUATING AN AI TOOL:

- Is FERPA/COPPA compliance mentioned?
- Is data used clearly and specifically?
- Can the tool be used without requiring personal information, such as complete addresses or phone numbers?
- Does it enable the complete deletion of student accounts and associated data?
- Is support contact information available?

When setting up accounts, use minimal data. Avoid full names—initials or nicknames usually suffice. Avoid using birthdays or photos unless necessary and always obtain permission beforehand. Many platforms support random usernames or class codes instead of email addresses. If emails are required, consider unique school-only accounts to avoid exposing personal details.

Data security becomes a daily habit—digital hygiene for your classroom. Regularly update shared account passwords and avoid sharing logins, unless essential. Always log out of shared devices and instruct students to treat their login details as securely as their house keys (never share or post them online). Remind everyone not to include sensitive information—such as home addresses or family matters—in chats, boards, or project comments unless necessary and adequately secured.

Build privacy habits into routines. For example, spend five minutes each week reviewing which apps are active and deleting unused accounts. Keep a simple list of tools in use and the information each collects; don't rely on memory months later.

Despite best practices, mistakes can happen. If sensitive data is accidentally shared—perhaps in an email or a public forum—remain calm and respond promptly. First, document the incident: note what was shared, where it occurred, and who witnessed it.

Use a Reporting Template:

- Date/time of incident.
- What was shared?
- Platform/tool involved.
- Names/IDs (if any)
- Immediate actions taken.
- Who was notified?

Report the incident to your school's IT leader or privacy officer quickly. They'll help contain the issue and determine the next steps, such as notifying families or changing passwords. Follow your district's data breach protocol.

Interactive Checklist: Your Digital Safety Routine

- ✓ Review privacy settings on all new tools.
- ✓ Use only school-approved platforms.
- ✓ Change passwords each grading period.
- ✓ Never post full names/photos publicly
- ✓ Remind students about safe sharing weekly.
- ✓ Maintain an updated list of all accounts/tools.
- ✓ Report suspicious incidents promptly.

AI Tools Aligned with Instructional Phases & Chapter 2

PHASE I: PLAN

Chapter 2. Building a Safe & Ethical AI Foundation

STRATEGY 1: SIMPLIFY LESSON PLANNING WITH AI TOOLS

AI TOOLS & SOFTWARE:

- Common Sense Education – AI Guidelines
  - **Description:** Guides educators and schools on ethical AI use with vetted resources.
  - **Link:** https://www.commonsense.org/education
- AI Ethics Curriculum by MIT Media Lab
  - **Description:** Curriculum for teaching students about AI and responsible use.
  - **Link:** https://aieducation.mit.edu/

Ultimately, protecting student privacy depends less on high-tech solutions and more on consistent routines and good sense. With AI, teachers act as digital guardians, fostering learning while diligently safeguarding student information.

CREATING AN ETHICAL AI USE POLICY—TEMPLATES FOR TEACHERS & SCHOOLS

Building an ethical AI use policy isn't just for tech specialists—it's for everyone in the school community, like teachers, students, parents, and administrators. Setting expectations up front reduces confusion, prevents mishaps, and maintains a safe classroom environment. A good ethical AI policy clarifies the roles of AI and humans, outlines their respective responsibilities, and establishes protocols for decision-making and addressing potential issues that may arise.

Transparency is essential: everyone should know how AI is used, what data it handles, and the intended outcomes, just like posted class rules. Accountability is crucial as well—there must be a straightforward way for students to voice concerns about AI-driven assignments and receive help if needed. Student autonomy matters: policies must protect students' rights to opt out or request alternatives without penalty.

Ready-to-use templates can simplify the process. For classroom teachers, a brief *"AI norms"* agreement can be very effective. **Here's a sample you can adapt:**

Classroom AI Norms Agreement

We'll always explain when and why an AI tool is used in class.

- No student will share their real name or personal information unless a parent or guardian approves.
- Anyone uncomfortable with a tech tool can speak up without judgment or penalty.
- We'll review AI tools together each semester for fairness and accuracy.
- Questions about AI are always welcome.

For your school or districts, a more comprehensive template might look like this:

School-Wide Ethical AI-Use Policy Template

- Purpose: Ensure all AI use supports student learning, safety, and equity.
- Scope: Applies to all students, staff, and digital tools used at or for the school.
- Transparency: Publish all AI tools used with clear summaries of what they do and what data they use.
- Student Choice: Students and guardians may request non-AI alternatives if feasible.
- Accountability: Staff monitor outcomes for bias, errors, or misuse and report concerns.
- Review Process: The Policy team meets at least once a year to review tools and update the policy.
- Contact Information: List of whom to contact for questions or issues.

Developing a policy is most effective through a team effort. Include teachers from various grades, administrators, tech staff, parents, and thoughtful students. Use real classroom scenarios to guide your discussions, e.g., "What if a reading app suggests something strange?" or "How do we support shy students with new tech?" Actual examples ensure policy relevance and improve buy-in.

Once drafted, share the policy widely—via staff meetings, family nights, or back-to-school packets—and encourage feedback from everyone, including anyone who spots gaps or confusing language. Clear instructions for offering feedback or suggesting changes help keep the document relevant and up to date. Effective policies evolve as users note new issues or improvements.

AI develops rapidly, so your policy shouldn't be forgotten in a file, it needs regular updates. Establish a review calendar (e.g., annually or at the start of each semester) to check for new tools or emerging concerns. Include all stakeholders in this review since last year's solutions may not meet this year's needs. Update the policy promptly when tools change features or privacy practices.

Assign clear checkpoints to support ongoing policy improvement: who collects feedback, how often tool use is reviewed, and how success or red flags are defined. Document these processes to ensure they happen reliably and celebrate improvements as they occur. If a student's concern is addressed or fairness is improved, share that progress to build trust and demonstrate commitment to better tech use.

Ultimately, an ethical AI policy is more than a documentation-guided daily classroom practice and empowers everyone to have a voice in how technology supports learning. As new tools appear, shared agreements help keep the focus on teaching actual students with real needs.

Communicating AI Ethics to Students and Parents—Scripts & FAQs

When you introduce a new AI-powered tool in your classroom, you can almost feel the ripple of curiosity—and sometimes a little nervousness—move through your students and their families.

A few might be excited, but others may worry about what this means for their privacy or how algorithms could impact fairness. To ease these concerns, transparent, honest communication makes all the difference.

Start by explaining things simply, the way you would to a neighbor who's not "into tech." I often begin with, **"AI is just a fancy way to say, 'computer helper.' These tools can help us find answers, give feedback, and suggest ideas. But they don't store or share your secrets with the world."** Direct, calm words set the stage for trust.

For parents, a straightforward letter works wonders. Something like:

"Dear Families, this year we will be exploring [AI Tool Name] in our learning activities. This tool helps us personalize lessons and saves your child's teacher time on routine tasks.

We have carefully reviewed its privacy safeguards. The tool does not sell or share your child's personal information with outside companies. Only school-approved staff can access data, and you are welcome to review our privacy practices at any time.

Please contact me with any questions or concerns."

Keeping your language concrete and avoiding tech jargon helps families feel informed rather than overwhelmed.

Students, especially those in upper elementary or middle school, benefit from a quick mini lesson that frames AI ethics in relatable terms.

Picture a whiteboard with two columns: "*What AI CAN do*" and *"What AI CAN'T do."* Ask students to brainstorm examples—maybe *"AI can check my spelling"* or *"AI can't tell if I'm having a tough day."* Then, talk through why fairness and privacy matter: ***"We use these tools to help everyone learn better, but we also make sure they treat everyone equally and keep our info private."*** Conclude with scenarios, e.g., ***"What if an app suggests something odd?"*** or ***"What if you don't feel comfortable sharing something online?"***—and talk through how to speak up if needed.

Families and students often raise similar questions year after year. Having answers ready saves you time and shows you take their concerns seriously.

Here are some common ones:

Will my child's data be sold or shared with strangers?

No. The tools we use are regularly reviewed for their privacy features. Your child's information is used solely for educational purposes within our school and is never sold to outside companies.

How does AI impact grading, and can it be perceived as unfair?

AI tools sometimes help with grading routine assignments. However, teachers constantly review results to ensure fairness. If something seems off or unfair, students can ask for a recheck.

Can I opt out of specific AI tools?

Yes. If you prefer not to use a specific tool, please let me know. We'll find an alternative that works for your child.

Will AI replace my child's teacher?

Not! AI assists by automating repetitive tasks, but teachers remain ultimately responsible for learning and decision-making.

When resistance arises—in an email from a skeptical parent or a raised hand during class —Listen closely before responding. Sometimes people just want reassurance that their worries are valid. If a parent says, ***"I'm worried about my child's privacy,"*** you can reply, ***"I understand your concern. I'm careful to use only trusted tools that protect student info."*** If a student asks if the computer is ***"spying"*** on them, clarify: ***"The tool only looks at schoolwork; it doesn't watch what you do outside class."***

--Maintain open communication--Not a one-time event: It's an ongoing process. Email updates work well for most families—short messages like:

"This month we're trying [AI Tool] to help with writing feedback. If you have questions about how it works or our privacy practices, just ask!" Newsletters can include a ***"Tech Corner"*** that explains new tools in kid-friendly language and offers ways for parents to try them at home if they're curious.

During parent-teacher conferences or open house nights, carve out a few minutes to discuss technology use. Share what tools are in play, how they help students, and what steps you're taking to keep everyone safe and respected. Some teachers find it helpful to have a printout of their FAQ readily available or an infographic illustrating how student information is protected.

For regular classroom updates, consider posting short reminders on your class website or learning management system: ***"Remember—never share passwords! Our math tool gives hints but doesn't collect private info."*** In class meetings, encourage kids to share their thoughts about digital fairness and privacy. When you show that you value their opinions and questions, they become partners in building an ethical tech environment.

These **practical scripts and routines** create a culture of trust and openness around AI in your classroom. Everyone, students, parents, teachers, knows what to expect and feels empowered to ask questions or raise concerns at any point along the way.

Spotting and Avoiding Bias in AI Tools —Red Flags and Review Rubrics

Bias in AI is a subtle yet significant issue. It creeps into digital tools in ways that can quietly shape student experiences, and sometimes, it takes a sharp eye to spot. Imagine you're using an AI-powered writing prompt generator, and you notice the tool seems to suggest stories about the same types of families, hobbies, or characters over and over.

Maybe it even describes only specific careers as **"for boys"** or **"for girls,"** or always sets stories in the same kinds of neighborhoods. That's bias at work—patterns in data that reflect old stereotypes. Or consider an AI grading tool that consistently scores essays from students who are English language learners lower than their native-speaking classmates, not because the ideas are weaker, but because of grammar quirks or nonstandard phrasing. These minor, repeated errors can make students feel unseen or unfairly judged.

Recognizing bias requires tuning into patterns. You might spot issues when certain groups, students with IEPs, those from different cultures, or kids who write in a unique voice, get less helpful feedback or have their responses flagged as "incorrect" more often. Some- times, bias can be hidden in the types of questions or vocabulary an AI platform uses.

If all the reading passages feature characters from just one background, or the math problems assume everyone goes skiing for winter break, you're looking at a lack of representation.

Even more subtle are moments when a tool "corrects" student work in a way that erases cultural expressions or unique dialects, sending the message that only one way of speaking is correct.

Vendors and tech companies aren't always transparent about how their AI was built or what data was used to train it. If you ask for details and get vague answers like "proprietary algorithm" or "improves learning outcomes for all students" without objective evidence, raise an eyebrow.

You want partners who show their work. Be cautious if a tool doesn't allow you to review or override its decisions, or if updates are rolled out without warning or explanation. Patterns of unexplained errors, mysterious grading changes, or limited customization options are big warning signs.

Evaluation of Potential AI Tools for Your Classroom

To help you evaluate potential AI tools for your classroom, I've put together a user-friendly review rubric.

This isn't about being perfect; it's about being vigilant and asking thoughtful questions. Here's what to look for:

- REPRESENTATION: Does the tool offer content that reflects the diverse cultures, languages, abilities, and experiences of your students? Are there multiple perspectives in stories, names, and images?
- FAIRNESS: Does the AI provide all students with equal opportunities to succeed? Does it avoid errors that punish students for language differences, disabilities, or unique ways of thinking?
- TRANSPARENCY: Does the company share information about how AI works? Were you told what data was used to train it? Can you see explanations for its decisions?
- ADAPTABILITY: Can teachers easily change settings, add content, or correct mistakes? Is there a way to appeal or override automated decisions?

Score each area from 1 (weak) to 4 (strong). If a tool averages below three across these categories, it's worth considering alternatives or pushing the vendor for improvements.

Bias In an AI Platform-Isn't Perfect: What To Do

Suppose you discover bias in an AI platform—maybe it keeps marking nonstandard English as wrong even when the meaning is clear. First, document specific examples with screenshots or student work samples.

Share these with your school's tech coordinator or administration.

When reaching out to vendors, keep your message clear and direct:

"I've noticed that your tool consistently flags responses written in [student's home language] as incorrect, even though the content is accurate. This impacts my students' confidence and learning. Can this be addressed?"

Maintain a record of all communication.

If the company ignores your concerns or fails to address the bias after repeated requests, advocate for your students by recommending that your school discontinue using the product. Propose alternatives that better serve diverse learners and share your findings with colleagues so they can make informed choices too. Schools should have a clear policy for suspending the use of tools found to be biased, students deserve nothing less.

Reflections: Bias, Recall & Discussion

Think back on any tech tool you've used—did it ever make you pause and wonder if everyone felt included? Jot down one example and what you did (or wish you'd done) about it. If nothing comes to mind yet, keep this question handy for your next lesson.

In closing, remember that building a safe and ethical AI foundation isn't just about checking boxes; it's about protecting student dignity and ensuring every voice counts.

"The rise of AI should mean the rise of human purpose."
— Demis Hassabis, CEO of DeepMind

WHY AI MUST ALIGN WITH INSTRUCTIONAL GOALS

Too many educators are being handed **"tech tools"** with no clarity on how they support student growth. Let's be clear: AI should never be used just because it's flashy. It must serve learning, not distract from it.

HERE'S HOW AI CAN HELP

(See tool descriptions in the next chapter and the Directory):

- Use ChatGPT to generate multiple-choice questions tied to the text.
- Use Diffit to rewrite passages at different reading levels.
- Use Curipod to create a student-led discussion with AI prompts on author intent

◆ CONNECT TO INDIVIDUAL GOALS AND STUDENT NEEDS

For Example:

- **For SPED students**: Use text-to-speech tools like Read & Write to allow access to grade-level content.
- **For ELL students:** Use Google Translate or Microsoft Immersive Reader for dual language scaffolding.
- **For Gifted learners:** Use Canva AI to challenge them to redesign a scene or debate the author's perspective

AI TOOLS—CLASSROOM SNAPSHOTS ◆

✓ Example 1: Math – 3rd Grade Fractions
Strategy: Explain the equivalence of fractions using visual models

AI Integration:

Use Brainingcamp to manipulate fraction bars.
Use Khanmigo to generate story problems at different levels of difficulty.
Use Edpuzzle to embed questions into fraction tutorial videos.

✓ Example 2: Social Studies – 6th Grade Government
Strategy: Compare democratic and authoritarian systems

AI Integration:

Use ChatGPT to model a mock debate.
Use Canva AI to create campaign posters.
Use Mote to allow students to record and respond to peer analysis.

◆ Alignment For Equity And Access

Alignment isn't just about standards. It's about reaching every learner.

- For Special Education
- Use speech-to-text for students with dysgraphia.
  Translate standards into student-friendly language using Diffit.
  Track IEP goal progress using AI-generated growth reports.
- For English Language Learners (ELL)
  Use dual-language tools (Immersive Reader, ReadTheory).
  Simplify vocabulary while maintaining core content.
  Offer pronunciation and listening practice via tools like ELSA Speak.
- For Gifted and Talented Students
  Offer extension prompts through AI, like

"What would happen if this law changed?"

****

Use research-based projects driven by AI suggestion engines.
Let students explore cross-disciplinary design using Canva or Soundtrap AI

◆ Mapping AI to Bloom's Taxonomy & AI Tool to Use Chart

Bloom's Level	Instructional Goal	AI Tool Match
Remember	Identify the main idea	Diffit summaries or Quizizz recall questions.
Understand	Explain the concept	Khanmigo tutoring dialogues.
Apply	Use concept in a new scenario	Curipod design activity.
Analyze	Compare/contrast	Edpuzzle with embedded. Socratic prompts
Evaluate	Judge author bias	ChatGPT debate simulation.
Create	Build digital products or story programming	Canva AI storytelling, or Scratch.

◆ The IEP Connection—Aligning To Personalized Goals

Let's walk through a real alignment example:
Student: Maria (SPED + ELL)
Goal: Improve reading fluency by 10 words per minute in 8 weeks
Standard: Identify the theme and summarize the story elements

AI Integration:
Use *text-to-speech* so she can hear passages first.
Use *Diffit* to scaffold texts at the right reading level.
Use *ReadTheory* to track weekly reading time and accuracy.
Use *ChatGPT* to help summarize using sentence frames.

With this alignment, Maria gains access and acceleration, without stigma or delay.

◆ Alignment Toolkit For Teachers

Checklist: Is My AI Tool Instructionally Aligned?

✅ Does it support the academic standard?
✅ Does it connect to a student's IEP or PLOF (Present Level of Functioning)?
✅ Is it accessible for all learning modalities (VAKT)?
✅ Does it promote equity, engagement, or deeper thinking?

📎 (Copy and paste this checklist for future self-checking)

◆ Reflection Tools

✍ Reflect:

What standard are you teaching this week that could be enhanced with AI?
Which group in your class could benefit from AI-powered scaffolding or extension?
What's one student's IEP or learning goal you could align to a tech tool today?

◆ Brain Teaser:

If you could design a "smart lesson planner," what would it do automatically for each student--Using this acrostic and the word ALIGNMENT to remember the essential steps, with (or without) AI,

🔤 ACROSTIC: ALIGNMENT

A = Align to standards
L = Leverage AI support
I = Include IEP needs
G = Guide based on data
N = Nurture all learners
M = Monitor progress
E = Engage with visuals
N = Navigate growth goals
T = Transform teaching

🎨 VISUAL TOOLKIT

Anchor Chart: "Start with Standards, Layer in Tools"
Poster: "Not All AI Is Aligned—But Aligned AI Is Powerful"
Template: AI Tool Alignment Log (free Google Sheet link in Appendix)

← CONCLUSION: AI USE IN SUMMARY

Integrating AI into instruction is not about bells and whistles. It's about clarity, intention, and alignment.

When done right, AI helps you:
Meet standards.
Reach IEPs
Empower every learner
Work smarter.
Teach with joy.

Lesson Plans: AI
For Effortless Lesson Planning

CHAPTER THREE

CHAPTER 3:

LESSON PLANS: AI FOR EFFORTLESS LESSON PLANNING

- ✓ **OBJECTIVE:** Teachers will create a complete AI-assisted lesson plan that integrates VAKT strategies and aligns with grade-level state and local standards.
- ✓ **VAKT ALIGNMENT TO CHAPTER & STRATEGY: VISUAL:** Slides and Infographics; Auditory: Ai-Generated Instructions; Kinesthetic: Interactive Brainstorms; Tactual: Printable Plan Templates.
- ✓ **BLOOM'S LEVEL:** Application, Synthesis, Evaluation
- ✓ **INCLUSIVE SUPPORTS:** Supports all learners by allowing teachers to plan for multiple modalities using AI.
- ✓ **USING AI LESSON GENERATORS**—Step-by-Step for Any Grade or Subject

Picture this: It's Sunday night, and you're staring at a blank lesson plan template, feeling stuck. Whether you're a new teacher anxious about tomorrow's science block, an experienced teacher trying to revitalize your math lessons, or a parent-educator needing ideas for another week of home learning, sometimes you just need a break from the copy-paste grind.

AI lesson generators step in here—not as magic wands, but as practical, time-saving assistants that help you create lessons tailored to your unique group of learners. Unlike random worksheet sites, these tools scaffold learning objectives, spark fresh ideas, and help you design activities for any subject or grade.

An AI lesson generator is essentially your digital co-planner, designed to understand curriculum needs and language. With tools like MagicSchool, ChatGPT with education plug-ins, Eduaide.ai, or Kuraplan, you can generate custom lesson frameworks in seconds. Yes, you heard me correctly. I said seconds- no more than a minute for non-tech-savvy persons, still a great time-saver.

You simply enter a prompt, such as "Create a 5th-grade science lesson on photosynthesis with hands-on activities," and the AI provides a lesson plan that includes objectives, essential questions, and formative assessments.

This means less time spent searching Pinterest or rewriting old lessons, giving you a foundational blueprint to adapt the lesson to all of your students' individual needs; thus freeing you for more creativity and connection with your students *(Explore this free AI tool for more details and practice sample lesson plans: www.MagicSchool.com)*

PHASE I: PLAN

Chapter 3. AI for Effortless Lesson Planning

STRATEGY 2. PERSONALIZE LEARNING PATHS FOR EVERY STUDENT.

AI TOOLS ALIGNED WITH CHAPTER, STRATEGY, AND PHASE

- Curipod
    - **Description:** Interactive slides from one-sentence prompts. Great for student engagement and planning personalized learning.
    - **Link:** https://curipod.cocham/
- **Diffit**
    - **Description**: Adapts any article or topic to various reading levels instantly.
    - **Link:** https://www.diffit.me/

- Eduaide.ai
  - **Description:** Personalizes lessons and quizzes with AI-driven suggestions.
  - **Link:** https://eduaide.ai/
- LessonPlans.ai
  - **Description:** Auto-generates standards-aligned lesson plans from simple prompts.
  - **Link:** https://www.lessonplans.ai/
- TeachMateAI
  - **Description:** Tool with 60+ generators to simplify teacher tasks.
  - **Link:** https://www.teachmateai.com/

Now, let's Get Started (NOTE: First, read the Disclaimer in the front of this book about being watchful for Free AI Tools, software, and Links)

To get started, open your chosen AI tool and find the lesson generator.
For Example, with MagicSchool, ChatGPT. Or Curipod

NOTE: You may need a plug-in, depending on your device. Use a laptop, if possible. Cell phones and iPads don't show all the necessary features.

--You'll find a prompt box at the top—your command center.

--Enter a specific prompt: *"Generate a third-5th math lesson on fractions that includes a visual demo and group game."*

--Hit enter, and you'll get a structured plan: Objectives like *"Students will identify and solve fractions in real-life situations, with 80%; 85%; 90%; 95%; 1--% proficiency—depending on your students' level of knowledge awareness (Bloom's critical thinking level)"*

--Create an engaging opener (like a "Fraction pizza party!"), guided practice, and a quick assessment. You can usually fine-tune the tone (formal, playful), select lesson length, and add other requirements such as vocabulary support or IEP modifications

--Read my Disclaimer in the front of my book about free software, and please adhere to it.

Customization is the key next step. AI does the bulk of the work, but it can't know your students' personalities or needs. Edit the output to include your favorite warm-ups or use language that fits your class culture. For instance, if your group enjoys movement, consider incorporating a kinesthetic activity, such as using blocks or jump ropes, to build fractions. If a learning objective is too broad, make it specific: swap "Understand plant life cycles" for "Describe the stages of a sunflower's life cycle with labeled drawings." Many platforms allow you to paste images or copy text for easy annotation in your planning document.

After customizing, review the plan for accuracy and relevance. AI can sometimes misinterpret standards or suggest content that isn't a good fit. Use this checklist: Does the lesson align with your goals? Is the tone age-appropriate? Are there any cultural assumptions or unusual phrasing that need correction? Double-check key facts ("Are penguins mammals?" They aren't!) and confirm the assessments truly measure what you taught. Revise anything that feels generic or off.

For example, suppose the AI produces this third-grade fractions lesson:

- Objective: *Students will identify and represent simple fractions using food.*
- Opener: *Class talk about sharing pizza.*
- Guided Practice: *Create fraction pizzas with construction paper.*
- Group Activity: *Fraction relay—teams assemble pizzas with different toppings.*
- Assessment: *Exit card with their favorite fraction pizza.*

A teacher might jot down notes: "Good opener—my kids love food! I'll use paper plates instead for more hands-on fun. The relay needs clarified rules and a timer. I'll add an oral check-in for my English learners to the assessment." This review ensures lessons are engaging and meaningful for every student.

Reflection Section: Try It Now!

--Pick an upcoming Grade and Topic.
--Enter a Prompt into your AI lesson generator,
--Then, review the Plan using the Checklist above.

What surprised you? What would you adjust to match your students' personalities? Note one idea to use next week and one improvement for the plans.

AI lesson generators don't silence your teaching voice—they amplify it, letting you spend more energy doing what matters most: building connections with your students, responding in the moment, and sparking excitement for learning.

Collaborating with AI for Standards Alignment—Ensuring Compliance

One of those not-so-glamorous realities of teaching is keeping up with standards. Whether it's Common Core, state benchmarks, or your district's checklist, every lesson you create is expected to line up with something official. If you've ever squinted at a spreadsheet, searching for the correct code to match your creative lesson idea, you know this is not the part of the job that sparks joy.

AI lesson planning tools are changing that chore. Now, instead of hunting through pages of standards or losing time cross-referencing, you can use AI to align lessons in just a few clicks.

These platforms aren't just innovative, they're organized, and they remember what you don't want to have to remember.

Here's How It Works:

When you open a tool like MagicSchool or Kuraplan, you'll often see a menu or filter labeled "standards," or "Basic" (meaning free, most times);

- Sometimes with dropdowns for grade, subject, or even specific frameworks like a state's Common Core skills to be mastered.
- You can type in your state or select the relevant organization, and the AI pulls up the standards linked to your topic.
- For example, if you're planning a sixth-grade social studies unit on ancient civilizations, you select your grade, and the tool pops up the applicable standards, sometimes even tagged within the lesson outline itself.
- This process is less about memorizing codes and more about focusing on student learning.
- After you've generated or mapped your lesson, AI makes it easy to document compliance.
- A built-in export or print option lets you save a standards-alignment report, which lists each lesson segment and its linked standard.
- You may receive a matrix that shows which objectives directly tie to expectations.
- This isn't just paperwork for the sake of paperwork. It helps when administrators ask for evidence during curriculum audits or when IEP meetings require proof of alignment for specific goals and related services (accommodations).
- You can file these reports digitally or attach them to lesson plans in your shared drive.
- No more last-minute scrambling when someone wants to see how your project-based math lesson addresses **"representing and interpreting data."**

Collaboration Can Be Key

Collaboration gets a serious upgrade with these tools as well. Suppose your team is preparing a cross-grade-level reading unit.

- Each teacher can input their standards into the shared AI planning tool.
- The platform tags every lesson part with its matching code—no more guessing if everyone's on the same page.
- Colleagues can leave comments or suggestions right in the lesson draft and refine together, all while seeing which benchmarks are covered (or missing).

If you are working in Google Docs, some AI plug-ins now add tags directly into each activity or assessment, creating a live map that everyone can reference.

Compliance Fears (A Personal Relevant Story):

Since I use to evaluate Superintendents of Title 1 Schools in Maryland when I worked for the Maryland State Department of Education, and as the Assistant Superintendent of Baltimore City School, I was in charged assisting the school system out of what is called a "consent decree", meaning the city was cited by the federal government's Office of Special Education Programs (OSEP) for non-compliance in serving special education students.

The federal government had placed the school system under the supervision of the mayor's office, who brought me in to solve the issue, since I worked for the MSDE evaluating the Superintendent, he deduced that if anyone could help with this situation, it was I.

OMGoodness! Where was AI when I needed it then? That was a monumental task that took me three years to work our way out of the non-compliance banner over the whole system's head and reputation. Well, with the help of the Almighty! (Praise God for Faith, Family, and Favor).

I developed several creative and innovative programs and processes that stirred the system around and instituted systemic processes to be proactive, interventive (like building a plane while you are flying it, sort of speak); and preventive--so as not to have that issue again, regardless of who is sitting in that seat.

My processes are not a bandage approach that lasts for a short while, then everyone goes back to doing the same old non-compliant tasks. It has to be systemic through my "3 x 3 x 3" Plan.

This AI Toolkit Book for My Teachers Took Priority

I began writing a book about my experiences and the systemic programs I developed and implemented over the years, including my tenure as Assistant Superintendent of the State Department, in foster homes for young men, and other areas. Stay Tuned, but I needed to get this book out first for my beloved teachers, who are working diligently trying to do it all—in other words, to do more with less and less.

Knowing my background and experience evaluating compliance issues vis-à-vis state and local systems, teachers in my grad class at Coppin State University, and others I have known over the years, often ask me about their compliance fears teaching special education students, Title 1 students, and other compliance-required populations:

"How do I prove my lessons meet every requirement?

What if someone audits my plans?"

AI-Generated Documentation Has Real Teeth Here.

- With each lesson tagged and archived, you can pull up evidence in seconds. --Administrators appreciate clear audit trails, especially when reviewing programs for SPED, ELL, or Title I students who require extra documentation.
- For example, an AI-generated alignment chart can be downloaded as a PDF, showing not only what is taught but also exactly which standard each objective addresses.
- This makes curriculum reviews much less stressful and helps justify instructional choices.

Another Bonus:

These reports aren't just for administrative peace of mind; they also help you identify gaps in your curriculum.

- If the matrix shows that persuasive writing isn't covered in your language arts block this quarter, you can plug that hole before anyone else notices.
- It's a bit like having a GPS for your teaching path—if you veer off course, you get a gentle nudge back toward your destination.

Administrators and team leaders find these AI-generated artifacts useful as they streamline reporting. Whether it's for year-end reviews, accreditation checks, or ARD meetings (Admission, Review, and Dismissal), regular meetings with detailed records mean less guesswork and more confidence that every box is checked. You can also use these exports as part of IEP or 504 plan meetings, showing families exactly how their child's instruction wraps around needed support and standards.

For teachers collaborating with multiple co-teachers or instructional aides, AI's tracking features help everyone stay coordinated. No more "Did we already teach that?" moments during planning meetings. You all see what's been covered and what needs attention, reducing overlap and keeping instruction focused.

Much of what we do as educators involves striking a balance between creativity and compliance, making learning meaningful while adhering to guidelines that ensure equity and consistency. AI isn't replacing our judgment; it's giving us clarity and time, so we spend less energy on cross-referencing codes and more on designing experiences that matter for kids.

And when someone from the district pops in with a clipboard and asks about benchmarks, you'll have every answer right at your fingertips—no panic required.

Customizing Ready-Made AI Templates for Diverse Learners

The truth is, even the fanciest AI lesson template won't magically fit your exact classroom right out of the box. Every group of kids brings its blend of strengths, quirks, and learning needs.

That's why I recommend starting by sourcing templates from trusted places, such as MagicSchool, OpenAI's Education Hub, or a district repository filled with teacher-tested gems.

Before grabbing a template, ask myself:

--Who built this?
--Who was it meant for, and
--How much am I allowed to change?

The best templates aren't just "plug-and-play"—they act as strong blueprints, waiting for your personal touch.

When you find a lesson that claims it's "differentiated," look closer.

--Does it offer built-in support for reading levels?
--Are there suggestions for hands-on or visual learners?

If a template reads like a generic worksheet and ignores language or ability differences, keep moving, even with some of the recommendations listed here. Although it is free at the time of publication, there is no guarantee that it won't be altered or restricted over time. Just be watchful.

Magic Happens in The Tailoring.

Once you've picked a solid starting point, the magic happens in the tailoring. Say you have a science lesson on photosynthesis, but your class includes English learners and students with IEPs. The original template might list vocabulary like "chlorophyll" and "carbon dioxide" with no support. I use a blue pen (I prefer blue to red—another one of my techniques) or a digital highlighter. Add visual elements, such as pictures of leaves, sunlight, and arrows, to illustrate the process.

For ELLs, I'll simplify definitions: "Chlorophyll is the green part of a plant." I might bold key words, use sentence frames ("Plants need ___ and ___ to make food"), or add a vocabulary matching game at the start.

For students who benefit from tactile activities, I'll swap written responses for hands-on alternatives—perhaps building a photosynthesis model with colored blocks or drawing a comic strip that illustrates each step.

Challenges, Not Just More Busy Work

Gifted students need challenges, not just more busy work. If I'm working from a math template about fractions, I'll toss in an extension question like, *"Can you invent a real-*

world problem where three different fractions all mean the same thing?" Sometimes I add an open-ended project—let them design their own fraction board game or research how fractions work in music composition. **For students** who need extra support, I'll break tasks into steps, add hints, or use color coding for clarity.

Speed matters on those mornings when you're scrambling. AI tools now include dropdown menus to adjust reading levels or language. Specific tools include translation features, which enable the printing of Spanish directions for parents and switching between English and simplified English for students as they develop their vocabulary.

Plugins can adjust instructions to be more visual or auditory. **One click,** and your instructions read aloud, or key points become icons instead of text blocks.

Let me show you a before-and-after with an example. The "**before**" template (straight from an AI bank) for a 4th-grade history lesson reads:

> **Objective: Students will describe daily life in Colonial America.**
> **Activity: Read textbook pages 102–109 and answer questions 1-5.**
> **Assessment: Short quiz on content.**

Now, Let's Transform It for A Mixed-Ability Class:

Objective: Students will compare daily life in Colonial America to modern life through the use of visuals and storytelling.

Activity: In small groups, students rotate through stations—one with primary source pictures (labeling items), one listening to an audio diary (with transcripts), one building a simple diorama using everyday materials.

Each group prepares a brief skit or comic strip showing differences between then and now.

Assessment:

Exit ticket where students draw or write one thing they learned at each station. ELLs may use sentence starters or pictorial responses. Gifted students can add a hypothetical element, such as ***"How would life be different if Colonial Americans had smartphones?"***

Notice how the generic plan expected everyone to read and answer questions while the revised version gives options—visuals, listening, building, and performing. It invites all learners in and provides room to stretch or scaffold as needed. The assessment flexes too: not just quizzes but art, writing, and speaking.

Here's another quick fix: A math template for multiplying decimals lists practice problems. For learners who struggle, I turn some problems into word stories— ***"If you buy 2.5 yards of ribbon at $1.60 per yard…"*** For those ready to take it further, I challenge them to create their problems involving decimals in recipes or budgets and explain their thinking aloud.

Customizing AI templates isn't about more work; it's about more innovative work. The right tweaks mean every student can access the core idea at their level. And honestly? Sometimes these edits spark ideas I never would have thought up on my own. When your classroom reflects this mix of support, challenges, and creative options, students notice—they feel seen and valued.

The next time you grab an AI-generated lesson, pause before printing or assigning it. Think about who's walking into your room that day: the kid who loves building things, the shy reader who needs pictures first, the math whiz itching for something more complex. Your tweaks transform a decent template into something alive—a lesson every learner can grab onto and make their own.

Timesaving "Plug-and-Play" AI Lesson Banks—How to Build & Share

Organizing the chaos of lesson plans, mainly when you teach multiple subjects or support diverse learners, can feel like herding cats on a windy day. That's where a personal or team-based AI lesson bank steps in as a lifesaver.

You can build a digital library to store AI-generated lessons that you find effective, rather than trying to come up with new ideas each time you plan.

Tags

Some teachers organize their folders by subject (such as "5th Grade Math" or "Middle School Science"). In contrast, others sort them by grade range or differentiation type, such as "SPED Supports," "ELL Modifications," or "Gifted Extensions." The trick is to make these folders easy to browse, so you're not digging through endless files while the clock ticks toward first period.

Curation matters as much as creation. When you add a new lesson to your bank, tag it with keywords: think "Fractions—Hands-on—3rd Grade—Visual Support" or "US History—Debate—7th Grade—Enrichment."

These tags help you or a teammate search quickly later. If your district uses standards-based tags, include codes like "CCSS.MATH.CONTENT.3.NF.A.1" or "NGSS 3-LS4-3."

Shared Spreadsheets or Databases

Many schools use shared spreadsheets or databases with columns for standards, themes, tech tools, and student group adaptations.

Keeping your lesson bank updated means setting aside time regularly to: -

--Review what's working,
--Replace outdated activities, and archive anything that didn't land well. If you've ever tried to reuse an old lesson and found it full of references to discontinued apps or dated pop culture, you know why regular clean-up is key.

Collaboration Transforms Good Lesson Banks Into Great Ones.

Picture a team of teachers building a shared Google Drive or OneDrive packed with AI-generated lessons.

Permissions enable everyone to add, edit, and comment, allowing the best ideas to rise to the top.

Maybe your school hosts a monthly **"Lesson Bank Jam"**—a relaxed after-school session with snacks where teachers showcase new AI lessons they've tried, swap feedback, and vote on which ones get added to the main collection.

This type of ongoing teamwork fosters a culture where sharing becomes second nature, and everyone benefits from fresh inspiration. It also prevents the dreaded **"reinventing the wheel"** problem—why should one teacher spend hours designing a differentiated reading unit when someone down the hall already has a fantastic template?

Best-Practice Guidelines

Best practice guidelines help everyone contribute effectively. For instance, your team might agree that every lesson in the bank should include teacher notes, standard tags, and at least one suggestion for modifying the activity for different learners. A shared **"read-me"** doc at the top of the drive explains folder structure, naming conventions, and version control—so nothing gets lost in digital limbo.

The impact of lesson banks on teacher time and morale is real. Take the case of Ms. Patel, a fourth-grade teacher who once spent six stressful hours planning each week. After her school embraced a shared AI lesson bank, her planning time dropped to two hours, and those two hours felt more focused and creative.

She could pull from a menu of ready-to-use lessons, tweak them as needed, and spend the rest of her prep time calling families or brainstorming enrichment projects instead of staring at a blinking cursor. Meanwhile, at another campus, the social studies department used their bank to standardize core units across three grade levels. This not only saved time but also ensured that students with IEPs or ELL needs received consistent support, regardless of which class they were assigned to.

CURATING LESSONS TOGETHER ALSO OPENS DOORS FOR FEEDBACK AND IMPROVEMENT.

Teachers start leaving comments like "Try adding a sentence frame for ELLs here," or "My students loved the group activity but needed clearer directions." Over time, the bank grows smarter, just like your teaching practice does when you share and reflect with colleagues.

The beauty of building and sharing AI lesson banks is that everyone wins:

New teachers get off to a strong start instead of floundering alone, veteran educators can refresh their practice without burning out, and students experience more consistent and engaging lessons across classrooms. It's not just about saving hours; it's about making every hour count for more.

A BONUS: SAMPLE & EDITABLE LESSON PLANS USING MY AI TEACHING & LEARNING FRAMEWORK

3 Completed Samples & 3 Editable Lesson Plan Formats--Based on My AI Teaching Framework To Reach EVERY Student

In Summary, Each Lesson Plan Format will:

- Follow the four teaching phases (PLAN, TEACH, ASSESS, REFLECT)
- Aligned with a specific Bloom's level and VAKT learning style.
- Include both a blank interactive version and a filled-in sample using AI tools.
- Include free AI links and visually marked cues (icons/colors) for VAKT differentiation.

The teacher can modify the lesson plan and fill in the blanks with other subjects to accommodate the specific needs of your students, as needed.

WATCH FOR:

The lesson plan format emphasizes both the teaching process (the 'how') and the content (the 'what'). Each sample lesson plan integrates the appropriate Bloom's levels of critical thinking, allowing teachers to see a completed one using AI for guidance and research analytics.

"Dr. Hattie N. Washington's Four-Phase Framework is a comprehensive approach to lesson planning that integrates clear objectives, evidence-based teaching methods, and artificial intelligence (AI) tools for continuous improvement." --Lamar Myers (AI & I, LLC)

Each lesson is structured into four phases – **Plan, Teach, Assess, Reflect** – incorporating **Bloom's Taxonomy levels and multisensory VAKT strategies (Visual, Auditory, Kinesthetic, Tactile) for diverse learning styles.**

The templates include visual icons for VAKT modes.

(e.g., 👁 Visual, 👂 Auditory, · Kinesthetic, ✋ Tactile)

Each sample lesson demonstrates how AI apps can streamline preparation and differentiation, helping teachers **"work smarter, not harder,"** in line with Dr. Washington's approach and teaching methods. AI makes the process significantly easier, more exciting, and a great time-saver, all while yielding immediate results of improvement and many smiles of self-pride.

LESSON PLAN 1:

All-Purpose Lesson Plan (Blank & Sample)

This template provides a **basic 4-phase outline** suitable for any core subject.

It emphasizes a **measurable objective** (so that each lesson's outcome is clear) and uses Bloom's six cognitive levels (Knowledge through Evaluation), as outlined on en.wikipedia.org, aligned with VAKT modalities.

The sample lesson demonstrates how a math teacher can incorporate AI-generated materials to facilitate differentiation and practice.

📝 BLANK TEMPLATE – "ALL-PURPOSE 4-PHASE PLAN"

PHASE 1 – PLAN

- **Measurable Objective:** Create specific, observable goals for student learning (aligned to Bloom's critical thinking levels and modalities).
- **PLOP (Present Level of Performance):** Baseline data on student skills/knowledge related to the topic (e.g., prior assessments, observations)
- **Formative Preparation Steps:** Materials, groupings, and warm-ups. (Include any pre-teaching or adaptive supports based on PLOP.)
- **Free Tool:** hslda.org.

Phase 2 – Teach

(Direct Instruction through Bloom's Taxonomy levels, with VAKT modes):

- Knowledge (Visual 👁): Introduce basic facts or concepts visually (e.g., slideshow, diagram, video).
- Comprehension (Auditory 👂): Explain and discuss concepts orally (e.g., mini-lecture, storytelling, Q&A).
- Application (Kinesthetic ·): Interactive activity to apply learning (e.g., lab, role-play, hands-on exercise).
- Analysis (Tactile ✋): Students analyze parts of the concept, possibly by touching/manipulating (e.g., sorting cards, annotating texts).
- Synthesis (Visual 👁): Creative task to synthesize ideas (e.g., drawing a concept map, designing a poster).
- Evaluation (Auditory 👂): Higher-order questioning or class discussion to evaluate understanding (e.g., students justify or debate).

Phase 3 – Assess

- Summative Assessment: Conclude with an activity or quiz to measure the objective (could be a written quiz, project, or presentation)
- Free Tool: prodigygame.com.
  - Criteria: Define how mastery is determined (e.g., 80%; 90%, 95%, 100% correct on quiz or rubric scores).

Phase 4 – Reflect 💡 AI.

- **Mastery Exercises (with Answer Key):** Follow-up practice for skill mastery. (Include an answer key for self-check or teacher use.)
- **AI Tools for Reinforcement & CQI:** Tech integration to reinforce learning or improve teaching. (e.g., AI-generated practice quizzes, data analytics to inform reteaching.)
- **Tool:** Send students using secure Microsoft Access, myemail.constantcontact.com, or via your school district-wide account.
- **Brain Teaser or Graphic Organizer:** Engaging extension (e.g., a riddle, puzzle, or organizer) to deepen comprehension and/ or recap or reinforce learning.

📌 Sample Lesson Plan –

Grade 5 Math: Multiplying Fractions

Phase 1 – Plan

- **Objective:** Students will **multiply two-place fractions** and simplify the result, demonstrating mastery with at least 9 out of 10 problems correct; Then 10 out of 10 with opportunity to redo with AI tools (individually and/or with a partner (a "friend").

- **PLOP**: Recent pre-test shows most students can multiply whole numbers but are less confident with fractions (e.g.,, adding fractions scores usually average 70%, at first; Then AI allows teachers to assign follow-up exercises and fun activities, until the student/s achieve greater scores-- (e.g.,, 80%; 90%; 95%; 100% correct on quiz scores).
- **One student with an IEP has accommodations for simplified reading.**
- **Formative Prep**: The Teacher gathers fraction manipulatives (fraction circles and strips) and a visual fraction chart.
- **Free Tools: MagicSchool**: The teacher can generate a warm-up of 3 review problems on fraction basics to activate prior knowledge (saving prep time by auto-creating differentiated questions)
- blaze. Today: Warm-up results will guide the lesson's emphasis.

PHASE 2 – TEACH

- **Knowledge (Visual 👁):** The teacher displays a **fraction multiplication example** on screen: $\frac{1}{2} \times \frac{3}{4}$:
  With an area model image. Students visually observe how one-half of a three-fourths area is shaded.
- **Comprehension (Auditory 👂):** The teacher explains the steps (multiply numerators, multiply denominators) and **discusses why** this works. Students ask questions and rephrase the rule ("half of three-fourths is three-eighths"). A quick whole-class Q&A ensures they *understand the idea behind the rule.*
- **Application (Kinesthetic ·):** Working in pairs, students use fraction circle cut-outs to *physically model* multiplication problems (e.g., find 2/3 of 3/5 by overlaying pieces). Moving and overlaying the pieces provides a hands-on way to apply the concept. Each pair then shares their result.
- **Analysis (Tactile ✋):** Students receive a set of **fraction word problem cards**. They sort each card into **"multiplication needed"** vs. **"not multiplication"** piles, justifying their choices. (For example, a card says, ***"You have 2/5 of a cake and give away 1/2 of that – how much of the cake is given?"*** which should go in the multiplication pile.) Manipulating and sorting the cards helps students analyze different types of problems. The teacher circulates to check the reasoning.
- **Synthesis (Visual 👁):** Using AI software (MagicSchool), each student creates a one-page **mini poster** showing a real-world scenario for fraction multiplication (e.g., a recipe: **"🍪 *If you have 1/2 of a batch of cookies and each person eats 1/4 of that, how much of the whole batch is eaten?"***). They draw the scenario and write the fraction equation with a solution. This creative task lets students synthesize the concept in a new context.

- Evaluation (Auditory 👂): Students share their mini posters in a quick "gallery walk" presentation. For each, the class or teacher asks an evaluative question (e.g., *"Why does your answer make sense?"* or *"How would the outcome change if we doubled the recipe?")*. Students must justify their reasoning, and the teacher listens for rational justification (or misconceptions -for further reinforcement with another fun activity AI tool).

PHASE 3 – ASSESS

Summative Assessment: To conclude, students take a 5-question quiz on multiplying fractions. It includes straightforward computations and a one-word problem. The quiz was auto generated using *MagicSchool's quiz tool,* tailored to this lesson's objective. For example, MagicSchool provided a mix of numerical problems and a context problem, which the teacher reviews and adjusts for clarity.

Criterion: Initial scoring: 7 or 8 out of 10 (70% to 80%); Then, eventually higher--10 out of 10--mastery.

- Grading: The teacher will grade the quiz and also evaluate each student's poster from the synthesis activity using a simple rubric: Clarity of process (How) and accuracy of math (What). Both will contribute to evaluating (critical thinking) whether the objective was met.

PHASE 4 – REFLECT ❢ *AI.*

- Mastery Exercises: For homework, students who didn't reach 80% get a targeted practice worksheet with similar fraction problems.
- The teacher used MagicSchool to AI generate a set of extra practice questions, problems, or exercises differentiated for each student's needs (e.g., visuals or simpler fractions for those struggling). An answer key is provided so they can self-check their work at home.
- AI Tools for Reinforcement & CQI: The teacher reviews quiz results by question. Using this data (and MagicSchool's item analysis feature, if available on your device), the teacher identifies that many missed the word problem.
- She plans a quick re-teach of word-problem strategies in the next class, demonstrating Continuous Quality Improvement by refining instruction based on assessment data.
  Additionally, the teacher explores Eduaide to generate a few enrichment word problems for advanced students (e.g., multi-step fraction problems), ensuring all learners are challenged today.
- Brain Teaser: To keep learning fun, the teacher ends with a fraction riddle: *"I am a number less than 1. If you multiply me by 2/3, you get 1/2. What fraction am I?"* Students ponder this brain teaser.
- This riddle sparks critical thinking and gives a sneak preview of using inverses, setting the stage for the next lesson.

- (Answer: $\frac{3}{4}$, Why? Since $\frac{3}{4}\times\frac{2}{3}=\frac{1}{2}$.)

LESSON PLAN 2:

Interactive Lesson Plan (Blank & Sample)

This template emphasizes **interactive teaching** through the use of live polls, quizzes, and collaborative activities.

It's ideal for science or social studies lessons where engagement and real-time feedback are key.

The filled-in sample shows a middle-school science lesson enhanced with an AI-powered interactive presentation tool.

📝 Blank Template – "Interactive & Multisensory Lesson"

Phase 1 – PLAN

- Objective: Define *content/skill goal.*
- PLOP: *Notes on students' current present level of knowledge (KWL) or misconceptions (from prior quiz/discussion).*
- Prep Steps: *Prepare interactive presentation slides, media, and any materials. (Consider tech setup: devices for students, internet, etc.)*

Phase 2 – TEACH

(Leverage interactive AI tech for engagement):

- Knowledge (Visual 👁): *Present background information with engaging visuals (e.g., slides with images, short video).*
- Comprehension (Auditory 👂): *Explain content* ***verbally; use think-pair-share*** *or* ***live poll questions*** *to check understanding.*
- Application (Kinesthetic ·): *Hands-on or movement activity (e.g., simulation, experiment, role-play) to apply the concept.*
- Analysis (Tactile ✋): *Students work with data or sorted information (possibly using manipulatives or digital drag-and-drop) to analyze deeper aspects.*
- Synthesis (Visual 👁): *Interactive creation – e.g., students collaboratively create a slide, drawing, or mind-map to synthesize learning.*
- Evaluation (Auditory 👂): *Students articulate conclusions or opinions (e.g., a group presentation or a guided class debate) to evaluate outcomes.*

PHASE 3 – ASSESS

- Summative Assessment: *Use an interactive quiz or project to assess learning (could be on an EdTech platform for instant results).*
- *Define success criteria.*

PHASE 4 – REFLECT ❢ AI.

- Mastery Exercises: *Assign any additional practice (such as gamified or interactive exercises) based on the quiz results.*
- *Provide answer explanations.*
- AI/CQI Tools: *Utilize AI for feedback and improvement – e.g., review analytics from an online quiz to inform next steps.*
- *Tool: myemail.constantcontact.com*
- *Or use an AI tutor chatbot for student practice.*
- Brain Teaser / Organizer: *Conclude with a fun quiz question, poll, or visual organizer (which could be completed through an interactive app) to reinforce the lesson in a novel way.*

"The future is not about man versus machine, but man with machine."
— Satya Nadella, CEO of Microsoft

SAMPLE LESSON PLAN –

Grade 7 Science: Photosynthesis

PHASE 1 – PLAN

- Objective: Students will **explain the process of photosynthesis** (how plants convert sunlight, carbon dioxide, and water into glucose and oxygen) and *illustrate* its importance in the oxygen-carbon cycle.
- PLOP: From a prior brainstorm, many students recall that *"plants make food," but misconceptions surfaced (e.g.,* some think plants "eat" soil). No formal knowledge of the chemical process has been acquired yet. The teacher plans to address these mis**conceptions explicitly.**
- Tool: Curipod: www.curipod.com
- **Prep Steps:** The teacher creates a slide deck using Curipod (an AI-powered interactive lesson tool). The deck includes a title slide, a simple diagram of photosynthesis, a multiple-choice poll, and an open-ended question slide.
- Curipod's AI "Do Magic" feature was used to automatically generate a draft quiz question and a drawing prompt via **teachertool.ai.**
- The teacher also queues up a short video clip showing time-lapse plant growth. Laptops or tablets are ready for students to respond to Curipod activities.

PHASE 2 – TEACH

- **Knowledge (Visual 👁):** The lesson begins with a bright infographic of the **photosynthesis formula** (CO_2 + H_2O + sunlight → glucose + O_2) displayed via the Curipod slide. The teacher plays a 1-minute time-lapse video of a plant growing, grabbing attention visually. Students *see* the inputs and outputs labeled on the diagram as the teacher briefly defines each term.
- **Comprehension (Auditory 👂):** The teacher verbally walks through the process: ***"Plants take in carbon dioxide from air and water from soil..."*** while students listen. She addresses the earlier misconception: ***"Notice, they don't 'eat' soil – the soil just holds water and minerals."*** After this explanation, the teacher launches a **Curipod live poll**: ***"Which factor do you think is most crucial for plant growth?"*** Options: Sunlight, Water, Carbon Dioxide, Minerals. Students vote on their devices, and the results are displayed instantly on the board. The class discusses the poll outcome – a great auditory dialogue correcting any misunderstandings (for example, if few chose sunlight, the teacher emphasizes its key role).
- **Application (Kinesthetic ·):** Students perform a quick **role-play simulation**: one student acts as the Sun (shining light), one as a Plant leaf, others carry cut-outs labeled CO_2, H_2O, O_2, glucose.
- In a fun physical routine, **"Sun"** tags **"Leaf"** with light; students carrying CO_2 and H_2O arrows move toward **"Leaf,"** and those with O_2 and glucose move away to signify outputs.
- This whole-body activity helps them ***experience*** the flow of the photosynthesis process. Laughter and movement reinforce the concept in memory.
- **Analysis (Tactile ✋):** The teacher distributes sets of **picture cards** (e.g., sun, root, leaf, water, etc.) and some unrelated items (e.g., **"moon"** or **"oxygen mask"**).
- In small groups, students arrange the relevant cards into the sequence of photosynthesis and exclude any distractors.
- They physically manipulate the cards on their tables, discussing what comes first.
- What isn't needed (e.g., they set **"moon"** aside as not part of photosynthesis).
- As they do this sorting task, the teacher circulates, asking probing questions like ***"Why did you leave out the moon?"*** to prompt **analysis.**
- Each group then briefly shares their final card sequence, explaining the role of each component (e.g., ***"We start with the sun because... without sunlight, the plant can't make food"***).
- **Synthesis (Visual 👁):** Next, the teacher uses **AI Curipod's collaborative whiteboard** feature to have students create a concept map.

- Each student adds one sticky note on the shared **digital board** stating *why photosynthesis is important*. **Some write "produces oxygen for animals,"** others "creates food for the plant." Together, they've built a visual web of ideas.
- Additionally, students **draw quick sketches** on the board (some doodle a tree, while others draw a person breathing) to connect their ideas.
- This real-time **collaborative** drawing **synthesizes** their understanding of how photosynthesis ties into the broader ecosystem.
- **Evaluation (Auditory 👂):** To wrap up the teaching phase, the teacher poses an **open-ended question via Curipod**: *"In your opinion, what would happen to life on Earth if photosynthesis stopped? Why?"* Students record short voice notes on their devices (or take turns speaking) to explain their evaluations.
- One student says, *"Without photosynthesis, no oxygen – animals and humans would die."*
- Another adds, *"No plants = no food for herbivores."* The class listens to a few responses, and the teacher guides a brief discussion, applauding the well-reasoned answers.
- This **evaluative** thinking (considering consequences) shows **deeper understanding (comprehension).**

PHASE 3 – ASSESS

- **Summative Assessment:** Students complete a 5-minute **quiz on Curipod** consisting of 3 multiple-choice questions and two short-response questions about photosynthesis. (Curipod's generator helped create these questions, which the teacher tweaked for accuracy, teachertool.ai.)
- For example, one question asks:
  -- *"Which of these is NOT needed for photosynthesis?*

  A) *Carbon dioxide*
  B) *Oxygen*
  C) *Sunlight*
  D) *Water."*

Another asks for one product of photosynthesis.
Since the quiz is digital, the teacher and students get instant feedback.
Criteria: The teacher considers **80% or higher** as mastery.

Students see their scores immediately, and the teacher's dashboard indicates that 20 out of 25 students have reached the mastery threshold – a quick gauge of class understanding.

PHASE 4 – REFLECT ❢ *AI.*

- **Mastery Exercises:** As an extension, students who scored below 80% are assigned an **interactive reinforcement**. The teacher uses **Eduaide** to generate a **personalized practice quiz** for each of these students.
- For instance, Eduaide provides simpler questions and immediate hints (like a guided tutorial) on photosynthesis.
- These students take the AI-generated practice at home, then check a provided **answer key with explanations** to learn from any mistakes.
- Meanwhile, students who showed mastery tackled a fun AI-generated **photosynthesis crossword puzzle** that the teacher found (clues include "gas plants release").
- **AI/CQI Tools:** The teacher **reviews** the AI self-grading Curipod quiz analytics, which shows that one question most often missed was about the **role of sunlight**.
- Noticing this trend, she plans to clarify that point immediately and reinforce it tomorrow before moving on, using ***continuous quality improvement (CQI) to adjust her teaching.***
- She also gets an automatic **item-by-item breakdown** from Curipod (AI-driven analytics are a huge time-saver).
- To **reinforce learning** for all, the teacher creates a **Curipod AI-powered revision game** (a short actual/false poll competition) for the start of the next class, targeting those tricky concepts.
- **Brain Teaser:** To end the lesson on a high note, the teacher uses a quick **Curipod word cloud** brain teaser:
- *"In one word, what do plants need most?"*
- As students submit their words via their phones, a word cloud forms on the board with **"SUN"** in big letters (most students answered **"sunlight")**.
- This **visual brain teaser** reinforces the key takeaway (sunlight) in a fun way. The class cheers when they see their words appear live.
- By integrating these AI tools and reflections, the teacher not only cements all of the students' learning mastery but also refines **her reflection practices for next time.**

LESSON PLAN 3:

Project-Based/Collaborative Lesson Plan (Blank & Sample)

This template is designed for **collaborative projects or inquiry-based lessons**, where students work together to create a product (poster, presentation, etc.).

It highlights the use of graphic organizers and creative tools, such as **Canva**. It is particularly well-suited for subjects like social studies or language arts, where students **synthesize** information into a project.

The sample demonstrates a history lesson that culminates in a student-designed infographic, utilizing AI for research and design assistance.

Tool: Canva: www.canva.com

📝 Blank Template – "Project & Collaboration Focus"

Phase 1 – Plan

- **Objective:** *Project outcome goal (e.g., "Students will research X and create Y to demonstrate…").*
- **PLOP:** *Note group skills, prior knowledge, or needed support. (E.g., "Students have practiced research in the library last week; one ELL student will need vocabulary support.")*
- **Preparation Steps:** *Assemble resources (texts, links), form student groups, and prepare any necessary graphic organizer templates or research guides. (Consider using AI to gather resource links or simplify texts.)*

Phase 2 – Teach

(Facilitate inquiry and teamwork):

- **Knowledge (Visual 👁):** *Introduce background info via visuals (timeline, map, etc.). Consider using a* ***Canva*** *infographic or chart to contextualize the content.*
- **Comprehension (Auditory 👂):** *Discuss key concepts as a class. Clarify project instructions verbally and answer questions.*
- **Application (Kinesthetic ·):** *Students engage in research or exploration. This could involve a gallery walk, handling primary source printouts, or moving between stations to gather information.*
- **Analysis (Tactile ✋):** *Provide a graphic organizer or note-taking matrix for students to sort and analyze information (fill in charts, categorize facts). They might physically arrange info (e.g., sticky notes on posters) to find patterns.*
- **Synthesis (Visual 👁):** *In groups, students create their project deliverables (posters, slideshows, etc.). They should incorporate visuals and text to represent their learning. (Leverage tools like* ***Canva's free templates*** *for quality designblog.tcea.org.)*
- **Evaluation (Auditory 👂):** *Groups present their work and explain their reasoning. Classmates and teachers provide oral feedback or ask questions (optionally using a peer-review rubric).*

PHASE 3 – ASSESS

- Summative Assessment: *The project itself serves as an assessment. Use rubric to evaluate content accuracy, creativity, collaboration, and other relevant aspects. Alternatively, or in addition, have a brief individual quiz or reflection to ensure that each student grasps the key concepts.*

PHASE 4 – REFLECT ❢ AI.

- Mastery Exercises: *If some objectives weren't met, assign individual follow-up (which could include a written reflection or additional questions). Provide answer keys or exemplars as needed.*
- AI Tools for Reinforcement & CQI: *Consider leveraging AI for feedback, such as using an AI grader to analyze common errors in projects or an AI chatbot to help students quiz themselves on facts.* ***The teacher reflects on project outcomes – perhaps using a survey (which could be AI-analyzed) to improve next time,*** myemail.constantcontact.com.
  Brain Teaser / Organizer: ***Use a fun extension to reinforce the project's content – Maybe a historical riddle, an "escape room" puzzle, or a collaborative mind map to connect this topic to the next.***

✦ SAMPLE LESSON PLAN

Grade 8 History: Causes of the American Revolution

PHASE 1 – PLAN

- **Objective:** Student teams will **research three major causes of the American Revolutionary War** and create an **infographic poster** explaining these causes, including at least one visual symbol or image for each reason. They will present their posters to the class.
- **PLOP:** Students have some familiarity with colonial America (they've learned about the 13 colonies). However, a pre-discussion revealed **gaps**: e.g., few could explain *"No taxation without representation."* One student (an English Language Learner) may need support with complex historical texts.
- **Prep Steps:** The teacher gathers *four* key event summaries (French & Indian War, Stamp Act, Boston Tea Party, Intolerable Acts) at reading levels appropriate for grade 8. Using Eduaide, the teacher quickly generates **simplified summaries** of these events for the ELL student (blaze.today), translating jargon into simpler English.
- **Tools:**
  Eduaide: www.eduaide.com; blaze. Today: https://blaze.today/blog/ai-lesson-plan-generators/#:~:text=Image
  Canva: www.canva.com (www.Blog.tcea.org)

- **A Graphic Organizer** is prepared for note-taking – a cause/effect chart for each event. The teacher also sets up student groups of 3 and creates a **Canva** class template for the infographic (with title **"Road to Revolution"** and some guiding text boxes) to save design time (blog.tcea.org). Each group will use that Canva template as a starting point.

PHASE 2 – TEACH

- **Knowledge (Visual 👁):** The teacher starts with a **timeline slide** showing key events from 1754 to 1775 (from the French & Indian War to Lexington & Concord). This visual timeline gives students an overview of the period.
- She also shows an iconic image (e.g., the Boston Tea Party engraving) to grab attention.
- As students look at these visuals, the teacher briefly defines **"Revolution"** and "**colony**" to ensure understanding.
- **Comprehension (Auditory 👂):** Next, the teacher describes **three major causes** in a story-like manner:
  (1) **Economic**: Britain's war debts → new taxes (Stamp Act);
  (2) **Political**: Colonists' anger over lack of representation;
  (3) **Social**: Events like the Boston Tea Party fueled resentment. Students listen and take part in a short discussion:
  "Why would paying taxes make the colonists so angry?"
- One student mentions that people don't like unfairness. The teacher clarifies with simple analogies (e.g., **"How would you feel if the school council made rules without letting your class vote?"**). She also outlines the upcoming ***project task*** verbally: each team will dive deeper into these causes and produce a poster. Questions are answered so everyone knows the plan.
- **Application (Kinesthetic ·):** The class rotates through **info stations** to gather details on each cause. Around the room, there are four stations. Station 1 features a short text on the Stamp Act accompanied by a copy of a stamp. Station 2 covers the Boston Tea Party with an image of colonists dumping tea, and so on.
- At each station, students do a quick activity: for the Stamp Act, they stamp a mock newspaper to feel what a tax stamp was; for the Boston Tea Party, they examine (replica) tea leaves and a chest image, etc.
- This mini gallery walk gets them moving and helps them tangibly connect with historical events. Each student jots one or two facts from each station in their notes.
- **Analysis (Tactile ✋):** Back at their tables, groups use a **cause-effect graphic organizer** to sort the information they gathered. The organizer has sections for ***"What happened?"*** and ***"Why did it upset colonists?"*** for each event. Students spread out printed fact cards (prepared by the teacher with key facts) and decide which cause each point belongs to.

- For Example, a card might read ***"Britain needed money after the war → Stamp Act (1765)"*** – Students place this in the economic cause section. They physically move cards under headings like Economic, Political, and Social causes.
- This tactile sorting requires them to analyze the nature of each cause. As groups fill in their organizers (writing notes and arranging cards), the teacher circulates, prompting deeper thinking: ***"Is the Boston Tea Party mainly a political protest or also an economic one?"*** Groups may debate and can place it under multiple categories if justified. By touching, reading, and organizing facts, students actively process the causes.
- **Synthesis (Visual 👁):** Now, each group synthesizes their findings by creating their **infographic poster**. Using the provided Canva template link, they collaborate on Chromebooks to input their text and choose visuals.
- For Example, one group titles their poster ***"No Taxation Without Representation!"*** and lists the **Stamp Act (1765)** as a cause, with a brief explanation and an image of a stamped document. Another section is titled **"Boston Tea Party (1773)"** and features an illustration of colonists at Griffin's Wharf.
- **Canva's library** provides free images (the teacher guided them to use license-free graphics), and the template's drag-and-drop interface makes it easy – even the ELL student can contribute by selecting pictures.
- *AI Assistance:* The teacher encourages groups to try **MagicSchool's**
- **"AI Research" tool** for a quick fact-check – one group uses it to verify a date, and MagicSchool confirms the **Intolerable Acts were in 1774**.
- Within 20 minutes, each team has a colorful poster merging text and imagery, demonstrating their synthesis of the historical causes.
  Tool: MagicSchool: www.magicschool.com
- **Evaluation (Auditory 👂):** Groups present their infographics to the class. As each team speaks, they explain *why* the events they listed contributed to the colonies' push toward rebellion.
- For instance, one group explains, ***"The Stamp Act angered colonists because they had no say – this was a political cause as well as economic."***
- The class and teacher ask a few questions: ***"Which cause do you think was the biggest catalyst?"*** – This prompts **evaluative thinking.** One group argues the **Intolerable Acts** were the tipping point; another insists it was the accumulation of many taxes.
- The teacher listens to their answers and uses a simple rubric to score each presentation (looking for accuracy of causes, clarity of explanation, and use of visuals).
- She also invites peer feedback: after each presentation, another group shares one positive comment **("The timeline you included was super clear!")** and asks one question.

- This presentation phase allows students to articulate **judgments** about the significance of each cause, thereby solidifying **their evaluative comprehension.**

PHASE 3 – ASSESS

- **Summative Assessment:** The **infographic poster and presentation** serve as the summative assessment for this lesson.
- **The teacher uses a rubric with the following criteria:** Content Accuracy (did they correctly explain the causes?),
- **Use of Evidence** (did they include key facts and dates?), Visual Clarity (is the poster well-organized and readable?), and Presentation Skills.
- Full points for **Content Accuracy** require all causes to be correctly explained without significant errors. Nearly all groups meet the expectations – their research and the graphic organizer work paid off.
- In addition, to ensure individual accountability, the teacher has each student submit a quick **exit ticket**: ***"Which cause of the Revolution do you think was most important, and why?"*** (This written reflection ensures every student, even those who were quieter in group work, can articulate their understanding.)
- The exit tickets show varied opinions but generally accurate reasoning, confirming that most students achieved the objective.

PHASE 4 – REFLECT AI.

- **Mastery Exercises:** The teacher identified two students who confused the event order, with one incorrectly stating that the Boston Tea Party led to the passage of the Stamp Act. For those students, she assigns a brief **timeline worksheet** on the events of the Revolution to reinforce the order.

- She found a free worksheet on Canva's educator resources and modified it for clarity blog.tcea.org.

- An **answer key** is provided so they can self-check. Meanwhile, students who demonstrated solid understanding move on to a short extension reading about **"The role of propaganda in the Revolution,"** which the teacher enriched with an AI-generated summary to keep it accessible.

- **AI Tools for Reinforcement & CQI:** The teacher uses an AI-driven assistant in Eduaide to **analyze common errors** in the posters.

  For instance, if the majority of groups omitted the French & Indian War as a cause, the teacher notes to emphasize that in future lessons (this insight was gathered by inputting project outlines into the AI for pattern recognition). Reflecting on the project, she realizes time management was an issue for one group.

To improve, she makes a note (in a **"Lesson Reflections"** Google Doc) to possibly assign roles in each group next time (researcher, designer, presenter) – an idea suggested by an **AI lesson planning blog** she read earlier.

This kind of reflection and adjustment embodies **CQI (Continuous Quality Improvement)** – using the experience data to tweak future instruction.

- Tool: myemail.constantcontact.com.
- **Brain Teaser / Organizer: Graphic organizers** can greatly enhance active learning. For example, the exit ticket in this lesson could be structured as a mini graphic organizer (**"Most important cause → Why?"**).

  In this lesson, as a fun closing challenge, the teacher gives a **historical riddle**: ***"I was a law that made colonists furious in 1765. What am I?"*** Students quickly answer, **"The Stamp Act!"** – a final recall that reinforces one key cause playfully.

  The teacher also displays a **word cloud** of all the causes covered, created using an online AI tool; the largest words are **"Taxation"** and **"Tea."**

  This visual recap and the riddle serve as both a brain teaser and a confidence booster, concluding the class with students feeling proud of their newly acquired knowledge.

Lesson Banks Are More Than File Folders

In Closing This Chapter 3, remember that lesson banks are more than file folders—they're **living collections powered by your creativity and community spirit.**

When you invest in building and sharing these resources, you're not just lightening your load; **you're helping your whole school grow stronger together.**

Next up: we'll explore how AI helps you **personalize learning paths,** so every student moves forward at their **own pace and style.**

"Any sufficiently advanced artificial intelligence is indistinguishable from magic."

— Inspired by Arthur C. Clarke's Third Law

Differentiating

With AI—Personalized Learning for Every Student

Chapter Four

CHAPTER 4:

DIFFERENTIATING WITH AI - PERSONALIZED LEARNING FOR EVERY STUDENT

- ✓ **OBJECTIVE:** Educators will design one differentiated activity using AI for three learning levels across a diverse classroom profile.
- ✓ **VAKT ALIGNMENT:** Visual: Leveled visuals and diagrams; Auditory: Text-to-speech tools; **KINESTHETIC:** Learning centers; Tactual: Personalized tactile resources.
- ✓ **BLOOM'S LEVEL:** Application, Analysis
- ✓ **INCLUSIVE SUPPORTS:** Ideal for Title I, SPED, ELL, and GT learners with scaffolded differentiation.

AI Tools & Software to Accompany & Reinforce

Phase II: TEACH

Chapter 4: Differentiating with AI—Personalized Learning for Every Student

Strategy 3. Automate Grading & Feedback-Personalize Learning Paths for Every Student.

- Khanmigo
    - **Description:** An AI tutor that helps students while teachers monitor progress.
    - **Link:** https://www.khanacademy.org/khan-labs
- Socratic
    - **Description:** Students take a photo of their homework to get explanations.
    - **Link:** https://socratic.org/
- ScribeSense
    - **Description:** An AI tool that helps grade handwritten or typed assignments.
    - **Link:** https://www.scribesense.com/
- ELSA Speak
    - **Description:** Helps ELL students improve speaking and pronunciation skills.
    - **Link:** https://www.elsaspeak.com/en/

AI-Driven Learner Profiles—Personalization from Day One

Suppose you've ever faced a classroom full of unique students. In that case, each with their quirks, abilities, and needs, you know it's nearly impossible to keep every critical detail organized, especially with new groups each year.

Tracking who needs extra support, who's passionate about dinosaurs, or who learns best solo quickly becomes overwhelming.

Instead of a mental jumble or sticky notes, AI-driven learner profiles offer a real solution: a dynamic, continually updated portrait of each student that equips you with actionable insights for differentiation, without adding to your to-do list.

What exactly is an AI-driven learner profile? Imagine a digital portfolio that does more than collect grades or snapshots; it learns and evolves with your students. These profiles are compiled from multiple data sources, including initial assessments, "about me" surveys, engagement patterns, as well as both teacher notes and student self-reflections.

AI connects the dots to spot trends that might slip past a busy teacher; you might receive a reminder that Maya excels with visuals or that Eli gets more engaged when history is connected to music.

This tech isn't just for gadget lovers—it's designed for everyday classrooms, recognizing the individuality of every student.

Getting These Profiles Rolling Is Straightforward.

In the first week, you can launch an AI-powered survey inviting students to share favorite topics, hobbies, and learning preferences. The best tools go further, letting you ask about learning methods ("reading, listening, building, moving?") and run quick skills checks to see where students start.

Adding baseline data is simple: flag which students have IEPs or 504 plans (while respecting privacy), include reading or math scores, and incorporate notes from earlier teachers or parents.

The AI then continually pulls in more data, including participation frequency, assignment completion speed, help requests, and behavioral patterns. The digital profile is dynamic, adapting as students change. Perhaps a shy student becomes more vocal in science, or an English learner seeks more help in a group.

These shifts are flagged for you, so you don't have to rely on just your memory. Your dashboard becomes your teaching ally, highlighting who needs scaffolds, enrichment, or peer support.

Grouping Students Becomes Much Smarter.

Rather than relying solely on instincts, you can build collaborative teams based on up-to-date strengths and interests, with AI making suggestions and encouraging rotation as students and data shift.

The system may notice that several students are highly engaged by animal stories, suggesting a natural group, or that pairing students needing language support with peer mentors could be helpful.

This also streamlines scaffolding and extension activities. The AI can recommend visual organizers for students flagged for language support or auto-assign audio instructions to auditory learners.

For advanced students, it suggests enrichment challenges, so differentiation requires less guesswork and extra prep.

Managing Data, Especially Sensitive Data

Managing all this data requires genuine care, especially regarding privacy and equity. Sensitive information—such as IEPs, home languages, and interests—should be securely stored and accessed only when supporting the student. Be transparent with families about the data collected and its use; this fosters trust, which keeps families engaged.

Also, check that the AI isn't oversimplifying or pigeonholing students—profiles should expand opportunities, not reinforce labels or stereotypes.

Interactive Reflection: Your Digital Learner Profile Checklist

- Launch first-day surveys that capture interests, learning preferences, and goals.
- Use AI recommendations for flexible grouping and targeted support.
- Securely include baseline reading and skill levels, as well as any necessary accommodations (e.g., IEP/504).
- Regularly update profiles with teacher notes and engagement data.
- Review your privacy settings monthly and involve your family in reviewing their data.
- Check for bias—are groupings too repetitive? Adjust as needed.

A dynamic learner profile transforms differentiation.

It eliminates paperwork and guesswork, allowing you to focus on relationship-building and seeing each student as a whole person, rather than just a name on a screen.

Adapting Content for SPED, Title 1, ELL, and Gifted Students with AI Tools

Differentiation used to mean spending hours after school tweaking lessons, rewriting worksheets, or hunting for "just right" reading passages. Now, AI gives you the power to instantly create versions of the same lesson for your varied learners, without marathon planning sessions. Suppose you're teaching a science concept about plant life cycles.

With a quick prompt, an AI tool can rewrite the original text at three reading levels. One version covers the basics, using simple sentences and visuals, perfect for SPED or emerging readers. Another uses grade-level vocabulary and adds a few comprehension questions for your core group.

The third version might incorporate advanced scientific terms and encourage independent research, making it ideal for your gifted students who crave a challenge. No more copy-pasting and rewording until midnight.

AI Accommodating EEL Students

Students who speak languages other than English at home benefit when instructions aren't a mystery. AI now enables the translation of assignment directions into dozens of languages within seconds.

--You copy your instructions into the tool,
--Select the required language (Spanish, Arabic, Vietnamese—whatever your class needs),
--A translated version is generated, ready to print or share digitally.

Benefit to Families: This accommodation enables families to support homework and gives ELL students confidence, rather than anxiety, when facing new assignments. Even for complex tasks, AI translation tools will allow you to review accuracy and make quick edits as needed.

Some platforms offer side-by-side directions in both English and the home language, allowing students to switch between them as they learn.

Scaffolding

Scaffolding accommodation has never been easier or quicker. Need a graphic organizer for a math concept? You tell the AI to create a visual chart showing "steps to solve word problems," and it generates a fill-in-the-blank organizer with icons for each step.

For students who require audio support, perhaps due to reading disabilities or temporary struggles, the same tool can convert any written passage into spoken word with just a click, allowing students to listen as they follow along. If you have learners with hearing challenges, AI now auto-captions video lessons nearly instantly, letting kids read what's spoken as they watch. Gone are the days when you'd have to wait weeks for someone in the district office to transcribe your videos or create visuals; now it's at your fingertips, ready whenever you need it.

Instructions themselves can be simplified without losing meaning. You paste your original directions into an AI platform and ask for a "step-by-step" version written at a fourth-grade level, or request bullet points with icons for each action. The system breaks it down cleanly, ideal for SPED, ELL, or anyone who gets lost in dense paragraphs. For students with organizational challenges or ADHD, you may want to prompt the AI to add checkboxes or reminders to help them keep track of each part of a project.

Gifted students often finish early and crave deeper exploration. With AI, you don't have to invent extension activities from scratch every week. You can prompt the tool: "Create three open-ended questions about renewable energy that require research and critical thinking." A set of prompts emerges, inviting students to debate which energy source is best for their city or design an experiment on solar power efficiency.

Run "Genius Hour" or independent study projects. The AI can generate project menus based on any topic—robotics, local history, art in science—offering ideas that push students beyond grade-level standards without leaving others behind. For enrichment in math, you might ask for tiered problem sets: basic practice at the top; real-world challenges at the bottom that require creativity and persistence.

Paperwork eats up valuable time, especially when you're tracking accommodation for multiple students on IEPs or 504s. AI-powered reporting tools now help organize your compliance documentation almost automatically. As you provide support—such as audio files, simplified texts, or translated materials—you log them into the system with a single click or a brief note.

The platform compiles everything into a running report: what was provided, when, for whom, and how often. When it's time for an IEP meeting or audit, you print or email a summary report showing exactly how supports were delivered. This not only saves hours but also helps you spot trends—maybe one student is using audio support daily but still struggling with comprehension, signaling a need for further adjustment.

Teaching in classrooms where no two students are quite alike can feel like spinning plates, hoping nothing falls. With AI's real-time content creation and documentation abilities, you go from improving every day to supporting each learner precisely where they are—quickly, efficiently, and without sacrificing your evenings or weekends.

Using Adaptive AI Platforms for Real-Time Learning Path Adjustments

If you've ever wished your lesson plans could shift to meet your class's immediate needs, adaptive AI platforms like **DreamBox, Smart Sparrow, and KidSense** are built for that. Unlike static programs, these tools adjust content in real-time: a student struggling with decimals receives simpler practice, while another ready for more advanced challenges gets tougher ones.

This cuts guesswork about who needs review and who can move ahead; the system identifies and nudges each learner along their personalized path.

Setting up an adaptive platform is straightforward. Teachers begin by accessing a dashboard to create accounts or import student lists. Most platforms start with a brief diagnostic—typically games, questions, or puzzles—to assess each student's strengths and areas for improvement.

The AI manages the scoring and pinpoints starting points for each learner. Teachers then set goals (individual or class-wide), like mastering multiplication facts by November. The dashboard organizes everything, and students log in to begin lessons that adapt as they answer questions, make mistakes, or demonstrate mastery.

As students progress, the platform collects real-time data. Progress bars and heat maps display mastered skills (green) and areas needing improvement (red), with each student's path unique.

For teachers, this **instant feedback is invaluable.** Instead of waiting for quizzes to spot issues, you see them live. Red zone clusters indicate where intervention is needed; long green streaks highlight students who are ready for more challenges.

For example, if half the class succeeds with multiplying fractions but struggles with word problems, you can target that for the next group lesson. If a student, such as Maria, excels, you might consider giving her an extension project or a peer mentoring role.

These platforms are ideal for blended or flipped classrooms. In station rotations, adaptive learning can be implemented as a single station, allowing for personalized, independent work while you guide other activities.

In a **flipped setup,** students do adaptive lessons at home, preparing for richer discussions or projects in class. The platform tracks progress so you know who's ready for the next step—no endless grading needed.

Good management comes from routines. Each week, check your dashboard for red and green zones. During planning, use the recommendations to guide small-group or individual reteaching, while others continue independently. Platforms let you easily adjust pacing or content if patterns emerge, and a quick mini lesson can address group struggles before moving on.

Student independence thrives in these environments. Learners receive instant feedback: correct answers are reinforced, mistakes prompt hints, and growth is visually tracked. Many get motivated by progress graphs or earning digital badges. It's common to hear excitement from students when they "level up," even in challenging subjects.

For families, adaptive platforms clarify progress. Parent reports highlight strengths and areas for growth in simple terms, making conferences collaborative instead of surprising. For administrators, the data shows student growth over time, guiding both classroom and schoolwide decisions.

No advanced tech skills are needed to start; most companies offer tutorials, and teachers find students adapt quickly after the first week. If your school uses multiple adaptive tools, rotate them by subject or day to provide variety and avoid digital fatigue.

When using platform recommendations, remember they are just data points. Your judgment still matters most. Sometimes, flagged skills may reflect distractions or rushed work, rather than a proper understanding.

Use these suggestions as launching points for talking with students: "You went through those geometry puzzles fast—were they too easy? Ready for a challenge?" or "Fractions were tricky yesterday—want to review together?"

Adaptive Learning Platforms

Adaptive learning platforms aren't about replacing teachers but amplifying their impact.

These tools:

- Provide real-time insights, and
- Allow you to spend less time guessing and more time addressing student needs promptly.

The classroom becomes more flexible and responsive, without the added stress of lesson planning or grading.

Case Studies:

Differentiation Success Stories Across Diverse Classrooms

The real power of AI in differentiation is clear in the firsthand experiences of teachers and students. In an urban middle school, Ms. Alvarez faced a significant reading gap in her sixth-grade English class, with over half of the students being English language learners and many of them being below grade level, including recent arrivals.

Her challenge: advancing everyone without losing top performers or leaving newcomers behind. AI-powered reading tools became essential. She used a platform that suggested customized texts and comprehension questions for each student, based on their most recent performance.

The system alerted her to insights that were easily overlooked, such as who flourished with visual support or who needed more vocabulary review. "I never thought my students would be so motivated by seeing their progress graphs," Ms. Alvarez noted.

These graphs inspired even reluctant readers to ask, "Can I see how much I've grown?" Far from being just numbers, they boosted confidence and sparked friendly competition.

By spring, the classroom's average reading scores rose nearly two grade levels, and parent conferences shifted from concern to celebration.

Across the county, in a rural elementary school, Mr. Carter managed a multi-grade class with several students on Individualized Education Programs (IEPs), ensuring that all paperwork was completed and their goals were met.

An AI-driven math platform helps by tracking progress and highlighting missed benchmarks in real time. It analyzed patterns, like a student excelling with hands-on activities on Fridays, or another making rapid progress after switching to oral responses. "The AI's recommendations helped me see patterns I'd missed before," Mr. Carter said, admitting past struggles with sorting through work samples and notes. The system also handled compliance documentation, making IEP meetings easier.

Students gained confidence from instant feedback and loved tracking their gains on dashboards.

One third grader proudly showed his chart during a parent meeting: ***"Look what I did this month!"*** By year's end, IEP goals were consistently met, and students previously anxious about math began requesting extra challenges.

In a suburban school near Atlanta, Ms. Kim balanced a classroom of gifted students, those with disabilities, and children from low-income families. She used AI for differentiation, mainly in a science unit.

The platform suggested **enrichment tasks** for gifted students—open-ended experiments and research prompts—and created simplified checklists for those needing more support. She saw students who were previously on the margins become group leaders, thanks to support tailored to their needs. "I was amazed at how quickly my class culture shifted," Ms. Kim said. Collaboration increased, with everyone working at their level.

A common thread among these educators is that **AI helped reveal blind spots and prompted new awareness.** Ms. Alvarez started paying more attention to social-emotional signals when the AI flagged a sudden drop in engagement. Mr. Carter found he could focus on creative instruction, less worried about paperwork. Ms. Kim realized that by trusting some AI recommendations, students felt noticed, and she didn't have to create endless custom materials.

Students themselves noted the impact. **One ELL student shared,** *"It feels good to see myself catching up,"* and a **gifted student said,** *"I like that my questions are harder now—I'm not bored anymore."* Such feedback illustrates how personalized instruction boosts motivation.

Practical takeaways: Start small—choose one subject or period to test an AI-powered tool. Use its insights as a foundation, not rigid instructions—your professional judgment still matters. Share progress data with students; seeing growth motivates them to set higher goals.

Source: *Personalized Learning: Source Gate Link (In Resource Directory).*

One of AI's key strengths is its ability to analyze data.

In education, this means tracking a student:

- Study habits
- Time spent on tasks.
- Quiz scores
- Areas of improvement

AI dashboards can present this information clearly to students, teachers, and even parents.

✓ With this data, decisions become smarter:

- Teachers can adjust lesson plans.
- Students can set personal goals.
- Parents can provide better support.

✓ AI-Powered Virtual Tutors and Chatbots

Not every student has access to a personal tutor, but AI can fill that gap. AI tutors are available 24/7. They can explain concepts, answer questions, and even administer quizzes to learners. These virtual assistants are beneficial for homework help and exam preparation.

✓ Popular platforms using AI tutors include:

- Duolingo (language learning)
- ScribeSense (assignment grading)

And let's not forget AI chatbots that help students find answers to questions, recommend resources, or even calm their test anxiety.

✓ Customized Learning Paths

Not all students want to follow the same path. Others prefer speed. Others need to slow down. AI helps by creating custom learning journeys.

✓ These AI systems consider:

- Skill level
- Past performance
- Interests and goals
- Learning style (visual, audio, reading, etc.)

For example, adaptive learning platforms adjust lessons in real-time based on a student's performance.

This approach leads to better understanding and long-term success.

✓ Inclusive Learning for All

AI is also making education more accessible for students with special needs.

- Text-to-speech tools help visually impaired students.
- Speech-to-text tools help those with writing difficulties.
- AI can translate content into different languages for non-native speakers.

This ensures that no student is left behind and everyone gets a fair chance of success.

✓ Benefits of AI in Personalized Learning

Let's take a step back and look at why all this matters. Here are some key benefits of AI in personalized education:

1. Better Student Engagement: Learning becomes more engaging when it's tailored to your individual needs. AI ensures students stay interested.
2. Improved Outcomes: With real-time feedback, custom lessons, and targeted support, students gain a deeper understanding and achieve higher scores.
3. Teacher Support: AI doesn't replace teachers; it helps them. Educators can focus on mentoring and creativity while AI manages the heavy lifting.
4. Time-Saving: Automated grading, content suggestions, and chatbots save time for both students and teachers.
5. Lifelong Learning: AI tools support not just school education but also professional development, online courses, and self-paced learning.

✓ Challenges and Considerations

While AI in education sounds excellent, it's not without its challenges.

1. Data Privacy: AI tools collect a lot of student data. It's essential to protect this data and use it responsibly.
2. Digital Divide: Not every student has access to the internet or the necessary devices for AI-powered learning.
3. Over-Reliance on Technology: Excessive reliance on AI can make students overly dependent on machines, rather than developing their problem-solving skills.
4. Bias in AI Algorithms: If AI is trained on biased data, it might give unfair results. Developers must be careful while designing educational tools.
5. Teacher Personalized Training: Although teachers do not require advanced technical skills to use most free AI tools and software, they still need some personalized training in AI.

   Otherwise, some tech features may be underused.

🚀 The Future of Personalized Learning with AI

The future looks bright—and smarter.

Here's what we can expect in the coming years:

- 🚀 From math to music, new apps will offer highly personalized learning in every subject.
- ☺ Emotion AI: Some tools will even detect a student's mood or frustration level and respond accordingly.

- ☺ Voice-Powered Learning: With voice assistants like Alexa or Siri, learning will become more interactive and natural.
- 👀 Gamified Learning: AI will make lessons more engaging, akin to games, to boost fun and motivation.
- 📕. Better Integration with Schools: Schools will partner with tech companies to integrate AI learning tools into classrooms.

FINAL THOUGHTS

AI is no longer just a buzzword—it's becoming the backbone of personalized learning. It helps teachers teach more effectively, students learn more efficiently, and parents stay more involved.

It's essential to use AI wisely. When managed correctly, AI creates a learning environment that is more engaging, inclusive, effective, and fun for all students.

If you're a student, explore AI tools to support your learning journey. If you're a teacher, try integrating AI into your classroom. And if you're a parent, encourage your child to embrace these innovative solutions.

The revolution in education is already underway, and AI is at the forefront.

📕 REFLECTION GUIDE:

- Consider your most diverse class this year. Where have you struggled to reach every student?
- Is there one area—reading, math, or writing—where instant feedback could help uncover hidden needs?
- Who could you invite to try out an AI tool—a colleague, student leader, or parent?
- ☺ Differentiation isn't about perfection or working alone—AI helps you reach more kids meaningfully and gives back precious time and peace of mind.

****

✅📕 In essence, progress is possible in any context, whether high need or high-achieving, urban, or rural, when your compassion and love for Teaching are combined with innovative technology.

As your classroom evolves, flexibility, openness, and willingness to learn will serve your students well.

Streamlining
Grading & Feedback with AI

Chapter Five

CHAPTER 5:

Streamlining Grading & Feedback with AI

- **OBJECTIVE:** Educators will use an AI grading tool to score and analyze results from a 10-question formative assessment.
- **VAKT ALIGNMENT:** Visual: Auto-generated reports; Auditory: Read-aloud explanations; Kinesthetic: Data sorting activities; Tactual: Printed assessment review sheets.
- **BLOOM'S LEVEL:** Knowledge, Evaluation
- **INCLUSIVE SUPPORTS:** Reduces the grading burden and provides immediate feedback for learners who need it quickly.

Automating Multiple-Choice and Short-Answer Grading with AI

If you've ever stared at a mountain of bubble sheets or a stack of short-answer quizzes and wondered if your eyeballs were about to mutiny, you're not alone. The grading grind can make even the most passionate educator question their life choices, especially after a long day when your brain is running on fumes and your to-do list is still a mile long. What if, instead, you could sip your favorite cup of tea, click a button, and watch as AI does the heavy lifting? That's not a pipe dream; it's happening right now in classrooms just like yours.

Let's break down how this works. Say you've just wrapped up a multiple-choice science quiz. Rather than manually checking every answer sheet, you can use **AI-powered platforms like Google Forms, GradeCam, or Edulastic to handle the grunt work.**

With **Google Forms**' quiz feature, once your students submit their responses online, the platform instantly applies your pre-set answer key and tallies scores across the board. You simply log in, view an organized summary, and spot trends within seconds—no more flipping through dog-eared pages or deciphering messy handwriting. For those of you still using trusty paper quizzes (hey, sometimes pencils just work!), GradeCam comes to the rescue. You scan student answer sheets with your phone or document camera, upload them, and the system reads each bubble, marks correct and incorrect answers, and even transfers results to your gradebook.

Short-answer grading used to mean deciphering creative spellings and a hundred ways to say **"photosynthesis."** Now, AI can scan student responses for synonyms, alternate spellings, and even creative phrasing. You upload responses into the platform, set up acceptable answer variations (think **"H2O,"** **"water,"** or **"dihydrogen monoxide"**), and the tool uses natural language processing to recognize correct answers, even if they're worded differently than your model key. This feature is a game changer for students who might know the right idea but don't always express it in textbook terms.

Integrating these tools with your existing learning management system (LMS) makes things even smoother. If you're using **Google Classroom, Canvas,** or **Schoology**, connecting **Edulastic** or **GradeCam** means that once the AI finishes grading, scores automatically sync to your digital gradebook.

No more duplicating effort or risking mistakes during late-night data entry. You assign the quiz through **Edulastic**; students complete it online or on paper. The moment the AI wraps up grading, every score is automatically entered into **Google Classroom's** gradebook fields. This seamless workflow means you spend less time moving data around and more time teaching—or maybe just catching your breath.

AI grading tools aren't just about speed; they're also about accuracy and fairness. Error detection features flag any answer that's ambiguous or nearly correct, so you can quickly review before finalizing grades.

For instance, if a student's short answer for "**What gas do plants take in during photosynthesis?**" is "**CO2**" but they misspell it as "**C02**" (with a zero), the AI highlights it for your attention instead of automatically scoring it wrong. This review step is a lifesaver for students who struggle with spelling or for ELLs who might phrase things differently. You can award partial credit or override automated decisions in seconds—no need to comb through every response manually.

AI Tools & Software for Chapter, & Phase

Phase II: TEACH

Phase III: ASSESS

Chapter 5. Streamlining Grading & Feedback With AI

Strategy 4. Differentiate Instruction

- Gradescope
  - **Description:** Grades exams and homework using AI to save teacher time.
  - **Link:** https://www.gradescope.com/
- Writable
  - **Description:** Provides structured feedback on student writing drafts.
  - **Link:** https://www.writable.com/
- Formative
  - **Description:** Creates quizzes and real-time insights from student answers.
  - **Link:** https://goformative.com/
- ZipGrade
  - **Description:** Scans and grades multiple-choice tests using a smartphone.
  - **Link:** https://www.zipgrade.com/

- Germini
  - **Description**: Germini is currently under construction but aims to create active learning experiences that encourage student engagement and participation.
  - **Use:** Intended for educators to facilitate dynamic interactions in the classroom, fostering a love of learning in every student.
  - **Link:** htps://germini.org
- Kahoot
  - **Description:** Kahoot! is a game-based learning platform that enables educators to create engaging quizzes and learning games to assess student understanding.
  - **Use:** Teachers use Kahoot! to engage students in learning through interactive quizzes and games that promote participation and collaboration.
  - **Link** https://kahoot.com
- Nearpod
  - **Description:** Nearpod is an interactive classroom tool designed to engage students through live polling, quizzes, and multimedia presentations.
  - **Use:** It helps teachers deliver high-quality instruction by promoting active learning and differentiation, allowing them to instantly see students' understanding through feedback.
  - **Link:** https://nearpod.com
- Pear Deck
  - **Description: Pear Deck** is an interactive presentation tool that empowers teachers to create engaging lessons with real-time feedback and student interaction.
  - **Use:** It is widely used to facilitate active learning and assess student understanding through collaborative activities and instant polling.
  - **Link;** https://peardeck.com
- Socrative
  - **Description: Socrative** is an assessment tool that allows teachers to create quizzes and engage students with activities that provide real-time feedback.
  - **Use:** It's utilized to monitor student progress and understanding instantly, making classroom management and feedback simpler and more efficient.
  - **Link:** **https://socrative.com**

- Edulastic
    - Description: Edulastic is an online assessment platform that enables teachers to create, administer, and analyze assessments while providing instant feedback to students.
    - Use: It is designed to align with standards and helps track student progress over time, making it easy to personalize learning tasks.
    - Link: https://edulastic.com
- Formative
    - Description: Formative is an assessment tool that enables teachers to create interactive assignments and quizzes for real-time tracking of student understanding.
    - Use: Teachers can use Formative to collect instant responses, assess student work in real-time, and adapt instruction on the fly.
    - Link: https://goformative.com
- Flipgrid
    - Description: Flipgrid is a video discussion platform that enables teachers and students to engage in discussions through video responses.
    - Use: It's widely used to promote student voice and collaboration, allowing learners to share ideas and perspectives through short videos.
    - Link; https://info.flipgrid.com

Here's where things get exciting: the time savings are real, not just marketing hype. Take Ms. Lee, a middle school teacher with 120 students. She used to spend two hours hunched over each set of math quizzes by hand. After switching to AI grading with Google Forms and GradeCam, she was able to finish grading in fifteen minutes per class. That's nearly two hours saved per quiz—hours she now spends calling parents with good news or brainstorming creative activities for her students instead of chasing red pens under her desk. According to recent surveys (see source in the APA list), teachers who use AI-based grading platforms report saving over 60% of their grading time on routine assessments.

Accuracy also gets a boost. Manual grading, especially after a long day, opens the door to mistakes—skipped questions, misread answers, or fatigue-induced errors. With AI handling calculations and flagging questionable responses for double-checking, grading becomes more reliable and consistent across large groups of students. Tech doesn't get tired or distracted by hallway announcements.

Reflection Section: Could AI Save Your Sunday Night?

Think about your last batch of quizzes or short-answer assignments. How much time did you spend grading by hand? Jot down your estimate and imagine cutting that time in half—or even by three-quarters—with AI help. What could you do with those extra hours each week? Write down three things personally or professionally you'd rather spend that reclaimed time on.

In the end, automating routine grading isn't about replacing your judgment—it's about giving you more freedom to focus on meaningful feedback and real teaching moments. Plus, your Sunday nights might start feeling a little less like homework detention and a lot more like well-earned rest.

Generating Personalized, Growth-Oriented AI Feedback for Students

Feedback is the heartbeat of learning. Do you know that moment when a student looks up, waiting for your reaction, hoping for more than just a number circled at the top of their paper? That's the power of feedback, and AI is now helping teachers deliver it faster, deeper, and more personally than ever before. Imagine students receiving immediate, tailored comments that don't just mark what they got right or wrong but also nudge them forward.

With AI tools like **Writable or Gradescope,** you can set up your feedback to match the tone and goals you want for your class if you're going to encourage risk-taking with gentle nudges— **"Great start! Try adding more detail about your main character's feelings."** Just pick the encouraging feedback mode. If you prefer clear, direct pointers, **"Your answer needs two pieces of evidence,"** that's easy to configure, too. You can even select constructive feedback for students who benefit from honest, actionable advice without sugarcoating.

Setting up these feedback parameters is simple. In Writable, for example, you tell the system what to focus on: thesis clarity, supporting evidence, voice, and mechanics. With a few clicks, you set the level of detail—do you want quick check-ins or in-depth paragraph responses? Once assignments are submitted, AI reads each one and applies your preferences, generating comments that match your teaching style and the needs of each student. In Gradescope, you can annotate directly on student work, but the platform also learns from your comments and can suggest similar responses for future submissions. This speeds up your workflow while maintaining consistent and meaningful feedback.

Growth-oriented feedback is where AI shines. Instead of stopping after marking something incorrect, these platforms suggest specific next steps that help students move forward. Picture a middle schooler who writes an essay with a weak argument. The AI highlights strengths—maybe their use of examples—then offers targeted tips: ***"To strengthen your claim, add one more fact from the reading."*** If a student struggles with reading comprehension, an AI-powered literacy tool might identify the pattern and offer a hint, such as, ***"Reread paragraph three and look for details about the setting."*** These prompts aren't generic; they're customized, helping students see exactly where to focus their efforts. Teachers can prompt the AI for revision ideas or extra hints by simply flagging areas of struggle in the dashboard.

One of my favorite features is how AI tracks feedback over time.

Imagine a dashboard that **not only shows grades but also visualizes which skills students are improving** and where they continue to struggle. Each assignment adds a new data point. You might notice that several students consistently miss inference questions in reading or struggle with transitions in writing. Instead of guessing or combing through piles of old quizzes, you get clear visual charts showing which skills are growing and which need reteaching. This helps you plan mini-lessons or adjust your pacing without having to fly blind.

However, **feedback is most effective when students are actively engaged** in the process. That's where AI-powered peer and self-assessment tools come into play. **Writable,** for instance, lets students review each other's writing using scaffolded prompts or sentence starters: *"One thing I liked was…"* or *"Next time, try…"* The AI guides them through providing balanced, specific comments, rather than just *"Good job!"* or *"Needs work."* This makes peer feedback less awkward and more useful, especially for younger students or those who struggle with confidence. **For self-reflection,** AI tools can prompt students to set goals based on past feedback *("This week I want to improve my topic sentences.")* and reflect on their progress after each assignment.

Teachers can see how students respond to different types of comments, whether they revise after encouragement versus direct correction, and use these insights to tweak their approach. Over time, this fosters a classroom culture where feedback isn't scary or punitive; it's an integral part of everyday learning. Students learn to anticipate suggestions and act on them, thereby building genuine ownership of their growth.

AI doesn't replace what you do best; it amplifies it. You still decide what matters most: which skills to prioritize, when to be tough, when to show grace. But with AI handling the heavy lifting, delivering instant comments, tracking patterns, and suggesting next steps, you have more time for face-to-face conversations and creative teaching moments that make all the difference.

Interactive Element: Try-It Reflection

After your next round of assignments, use an AI tool to generate feedback for one student. Review the suggested comments and tweak them as needed to match your voice. Share the feedback with your students and ask them how it feels, does it help them know what to do next? Jot down their reaction in your planner for future reference.

AI Tools for Rubric-Based Grading—How to Set Up and Use Them

Rubric-based grading is a lifeline for teachers who want to give transparent, fair, and meaningful feedback on essays, projects, and presentations without losing their sanity or evenings. AI-powered platforms like **Turnitin, Gradescope, and Peergrade** now make this grading method not just possible but refreshingly manageable. Imagine uploading your custom rubric—those columns for **"organization," "content,"**

"mechanics," and "creativity"—directly into the platform—no more juggling paper rubrics or opening a dozen browser tabs to check the standards.

Instead, you click, drag, and define criteria right on your screen. For example, when setting up a writing rubric for a high school literary analysis, you can add **"thesis clarity," "textual evidence,"** and **"original insight"** as categories. You decide the weight of each section and save it as a template for future assignments or to share with colleagues.

Once your rubric is uploaded, the AI takes over the initial scoring by analyzing student work through the lens of your criteria. If you're grading a social studies debate, the system examines each submission for argument strength, appropriate use of evidence, logical flow, and even creativity in presentation.

The AI doesn't just tally spelling errors or count paragraphs; it reads for content, checks for logical structure, and flags areas where a student's reasoning is strong or weak. Each criterion is scored according to the standards you set. For instance, a student who backs up every claim with historical examples gets top marks for "evidence use," while another who wanders off-topic is gently nudged to tighten their argument.

The beauty of these AI platforms is that you're never handing over total control. After the AI applies preliminary scores and generates comments based on rubric descriptors, you can review every flagged case—those messy middle-ground essays that could swing high or low depending on nuance. **You read the AI's notes,** see where it hesitated ("argument strength: borderline between 'good' and 'excellent'"), and **make adjustments** as your professional judgment dictates. Maybe you spotted a clever insight that the machine missed, or perhaps you realize that the student's tone was just right for persuasive writing, even if not perfectly formal. This human calibration ensures that grading remains fair and responsive to the actual effort of students.

Formative assessment becomes a breeze when AI works hand-in-hand with your rubric. Instead of waiting weeks to spot trends, you get almost instant feedback loops: after an essay unit, the platform compiles results into easy-to-read visuals. Heatmaps show how each student—and the class as a whole—performed across every rubric category. Red zones highlight common trouble spots, such as "weak transitions" or "missing citations," while greens signal strengths in areas like "voice" or "evidence integration." These analytics don't just sit on your computer; they help you plan tomorrow's lessons. If half your students missed points in "argument strength," you know to reteach persuasive techniques before the next assignment.

Grouping students for targeted support is no longer a matter of guesswork. With a glance at rubric data, you can match learners who struggle with structure together for a mini-lesson or pair those with strong creativity but weak mechanics for peer review. The goal isn't just to identify weaknesses but to leverage strengths across your class community. Rubric analytics even allow you to refine your rubric over time. Notice that everyone is scoring low in a particular area. It may be time to revisit how that skill is taught or adjust expectations to match developmental levels better.

A feature that often gets overlooked is the ability to customize comments for each criterion. Instead of generic feedback like ***"needs more detail,"*** AI can generate specific suggestions: ***"Try adding two more quotes from chapter five to support your claim about the main character."*** You can refine these comments before sending them to students, ensuring your voice remains distinct even as AI handles the bulk of the work. Over time, these platforms learn from your adjustments, improving their future recommendations to better align with your teaching style and classroom culture.

AI doesn't just speed up grading, it gives teachers more control and sharper insight into student growth. You spend less time sorting through papers and more time building relationships, coaching the next steps, and celebrating small wins with your learners.

For students, getting clear feedback on exactly where they hit the mark (and where they didn't) makes revision feel doable, not mysterious, or impossible. **And for parents or administrators peeking in,** rubric-based AI grading provides **transparency**: everyone sees exactly what was measured and how scores were awarded.

Rubric-based AI tools are changing from a solo marathon into a team effort between teacher intuition and innovative technology. The result? More energy for creative teaching and fewer late nights spent buried under piles of essays.

Addressing Grading Bias and Ensuring Equity with AI Analytics

Fairness in grading isn't just a buzzword; it's something every student, family, and educator deserves. Even the best teachers, with all the right intentions, can fall into unconscious bias. Maybe it's a soft spot for a student who reminds you of your younger self, or perhaps a tough week clouds your patience. Now, with AI analytics, you have a partner that can help you spot these blind spots before they quietly shape your classroom outcomes. Imagine, instead of relying purely on gut feelings or occasional gradebook reviews, you have real-time dashboards that highlight patterns you'd never notice on your own.

AI analytics sift through piles of grading data in seconds and reveal disparities that might otherwise slip through the cracks. For example, after running a batch of essays through your platform, you might see a dashboard highlighting that English learners are scoring lower on argument structure, or that boys and girls are getting uneven marks on creativity. These equity reports don't just show numbers—they spotlight where unconscious bias might have crept in, or where an assignment or rubric favored one background over another. You'll see color-coded graphs and charts breaking down average scores by demographic groups. Suppose gifted students consistently score higher on open-ended projects while SPED students lag. In that case, that's a signal to dig deeper, not to blame anyone, but to open up a conversation about what support or changes are needed.

Calibrating AI grading tools for fairness starts with an intentional setup. When you train your platform to grade assignments using rubrics, it's wise to upload sample responses from across your classroom's diversity: essays from ELL students who use inventive grammar, SPED students who organize thoughts in unconventional ways, and gifted students whose writing may leap beyond grade level. This gives the AI a broader view of what quality looks like for different learners, rather than only rewarding one style. You can adjust sensitivity settings, allowing the AI to accommodate creative approaches and avoid penalizing kids for minor language differences or unique interpretations. If you notice the machine is consistently harsher on certain types of responses, tweak the calibration, add more sample answers, or even bring in colleagues to help check for fairness.

Once your platform is running equity audits, it's time to roll up your sleeves and **interpret the reports.** Most AI tools provide **visual dashboards that summarize grading data by gender, language background, special education status, and other factors.** Review these regularly—weekly or monthly—to spot trends. Perhaps the dashboard indicates that students receiving reading accommodations aren't making the same gains as their peers on written responses. **Is the feedback language too complex? Are instructions accessible?** Then, **Use these findings** as springboards for team discussions or professional development. You might decide to **revise prompts, simplify rubrics, or collaborate with ELL or SPED specialists** to make feedback more **actionable for all students.** Equity dashboards aren't about catching someone in a mistake; they're about building a bridge toward better practice and stronger outcomes.

Real-world stories demonstrate the profound impact of these changes. Take a **Title I school I worked with in Baltimore.** After implementing AI-driven bias checks, teachers discovered that their scoring patterns unintentionally favored students who wrote in standard English and penalized others for minor language quirks. By adding more diverse sample responses for calibration and adjusting the rubric criteria ***("skillet" vs "frying pan", "purse" vs "pocketbook",*** etc.), they saw a 30% reduction in grading disparities within one semester. Teachers felt more confident that every student was being seen **for their thinking—not just their grammar or background—** and families also noticed the change. The conversation shifted from "***Why did my child get marked down again?"*** to ***"I see exactly where my child needs help—and what progress looks like."***

The ripple effect of using AI analytics for equity extends beyond numbers; it fosters trust and transparency with students and their families. When you share how you're working to remove bias—maybe at a parent-teacher meeting or in a student conference—you invite others into the process. Students learn to advocate for themselves if they feel feedback doesn't match their effort. Parents know their child's strengths won't be overlooked because of language or different learning needs.

These tools aren't about chasing perfection but about getting better every day. They help teachers reflect honestly, adjust quickly, and celebrate when progress is made.

Bias will always try to sneak in—AI just gives us better glasses to see it and the tools to push back.

In closing this chapter, it's clear that when we **combine teacher care with innovative technology, equity moves from an idea to an everyday reality.** The next step? Exploring how AI can not only assess but also actively inspire creativity and engagement across every learner, no matter their starting point.

Onward to sparking curiosity and joy in your classroom!

"AI doesn't replace what you do best; it amplifies it."

—Dr. Hattie N. Washington

Creative Engagement:
Supercharging Student Motivation with AI

CHAPTER SIX

CHAPTER 6:

CREATIVE ENGAGEMENT—SUPERCHARGING STUDENT MOTIVATION WITH AI

- ✓ **OBJECTIVE:** Students will utilize one AI tool to create a multimedia presentation that expresses their understanding of a lesson topic.
- ✓ **VAKT ALIGNMENT:** Visual: Videos and graphics; Auditory: Narration and music; Kinesthetic: Creation of digital stories; Tactual: Printed storyboard planners.
- ✓ **BLOOM'S LEVEL:** Synthesis, Application
- ✓ **INCLUSIVE SUPPORTS:** Engages all learners with expressive freedom and tech-based creation options.

LEVERAGING AI FOR GAMIFIED LEARNING EXPERIENCES–

Step-by-Step Guides

If you've ever watched students light up when a lesson turns into a game, you know that engagement can transform even the most reluctant learner. I remember the first time I saw a whole-class review morph from sleepy heads to cheers and high-fives, just because I turned questions into a digital challenge.

Now, with AI-powered gamification, that energy isn't just reserved for special occasions—it can become a daily part of learning, tailored to fit every student's pace and style. Imagine a classroom where students ask more complex questions, where ELL learners see their names rise on a leaderboard, and where even the shyest child beams as a team captain. That's not wishful thinking, it's what happens when you put AI-driven game mechanics to work.

****

AI TOOLS & SOFTWARE FOR CONTENT:

PHASE II: TEACH

Chapter 6. Creative Engagement—Supercharging Student Motivation with AI

STRATEGY 5. ENGAGE STUDENTS CREATIVELY—BRING SOME SPARK BACK TO LEARNING.

- Book Creator with AI
  - Description: AI-enhanced tool for students to create digital books.
  - Link: https://bookcreator.com/
- **Storybird**
  - **Description:** Let students write and illustrate AI-enhanced stories.
  - **Link:** https://storybird.com/
- **Animoto**
  - **Description:** Turn lessons into engaging videos with drag-and-drop AI.
  - **Link:** https://animoto.com/
- **Tome AI**
  - **Description**: Storytelling platform enhanced with AI design elements.
  - **Link:** https://tome.app/

AI-enabled platforms like **Classcraft** and **Quizalize** make it simple to turn dry content into something that feels like an a**dventure.** You start with your regular quiz questions—maybe vocabulary words, math problems, or science facts. Instead of handing out paper or clicking through a boring slideshow, you import these questions into the platform.

Instantly, the system transforms them into **quests, challenges, or competitions. Classcraft,** for example, allows you to **assign point values to each question;** students **earn "XP" (experience points)** not only for correct answers but also for demonstrating grit or teamwork.

Quizalize takes your set of questions and builds them into a **game** where students **compete individually or in teams,** earning badges and climbing levels as they progress. **The best part?** *You don't have to be a gamer yourself—the setup is drag-and-drop simple.*

Personalization is where AI takes gamification to the next level. Instead of a one-size-fits-all scoreboard, these platforms use student data to adjust the experience. Let's say you have a student who consistently struggles with multiplication facts; after a few rounds, the AI recognizes this pattern and introduces extra practice disguised as bonus rounds or "side quests." Meanwhile, the student who's breezing through gets **"level-up"** challenges that push them into word problems or multi-step tasks. Badge systems can be set up so that each learner is rewarded for growth, not just speed—perhaps an ELL student earns a "Language Legend" badge for improving their English vocabulary. In contrast, a SPED student receives a "Perseverance Pro" badge for sticking with tough questions.

Setting up adaptive points and badge systems is straightforward. In Classcraft, you build profiles that reflect your class's diversity—reading accommodations, language supports, or enrichment needs can be built right in. Quizalize offers dynamic leaderboards that highlight individual growth and team effort rather than just raw scores. **Teachers can adjust settings so that badges appear for different milestones,** such as mastering ten new words, completing a round without hints, or demonstrating collaboration in a group challenge.

Formative assessment becomes seamless within this system. As students play, the AI tracks their responses in real time—who needs more practice with fractions? Who's mastering spelling but missing context clues? Platforms like Kahoot! Now, we offer AI-powered question recommendations. After analyzing how students perform on one set of questions, the platform suggests new ones that target weak spots or introduce more complex thinking. Immediate feedback— ***"Great job! Try this next level!"***—comes with every attempt, letting students know where they stand and what to work on next. **For teachers, colorful dashboards display mastery by topic and student subgroup, allowing them to identify trends or gaps at a glance.**

Of course, tech only works if everyone has access and feels included. **Equity matters here as much as anywhere else in education. If your classroom doesn't have enough devices for each student, consider rotating through stations or using team formats where one device is shared per group.** Don't forget about your **shy students; rotate team captains so everyone gets a turn in the spotlight, not just the loudest voices.** Some platforms let you assign anonymous avatars, which helps students who feel anxious about public scores participate without fear.

Gamification For Creativity

Gamification isn't just about keeping score; it's about building motivation and community. When used thoughtfully, it helps all students, including those who often get left behind, feel seen and celebrated for their progress. If you ever notice sure students hanging back or becoming discouraged by competitive elements, consider adjusting the rules: offering private achievement badges or allowing self-paced quests where learners can level up on their terms.

Interactive Reflection: Building Your First AI Game Quest

Think of one lesson this week that feels flat—maybe vocabulary review or a math skills check. Jot down five core questions or problems you want students to master. Log into Quizalize or Classcraft (or another gamification tool your school supports) and upload those questions. Explore the settings: Can you assign different badges for creativity or improvement? Will you let students choose avatars?

After running your first game, reflect on who surprised you with their engagement. What tweaks would make it even more inclusive next time? Write down one adjustment for your next round—maybe rotating roles or setting up private challenges for students who need a confidence boost.

With AI-powered gamification, learning becomes something everyone looks forward to—even on Mondays.

Creating Interactive AI-Driven Simulations and Virtual Labs

Nothing beats watching a student's eyes widen as a difficult idea suddenly comes alive. AI-powered simulations and virtual labs take what used to be abstract—like the way genes mix, the magic behind chemical reactions, or how economies grow and crash—and make it something you can click, drag, observe, and even mess up without fear. Instead of memorizing facts about mitosis or the water cycle, your students can see cells split in real time, adjust rainfall in a digital ecosystem, or run a business in a virtual town.

These tools don't just boost engagement. They level the playing field for students who struggle with reading dense text, for English language learners who need visual cues, and for those who crave hands-on learning but don't always thrive with pencil and paperwork. Virtual science labs, such as Labster or ChemCollective, allow you to assign experiments that many schools could never conduct in person—think DNA extraction or physics challenges—without worrying about broken beakers or running out of supplies.

Setting up one of these simulations is much less intimidating than you might expect. Suppose you're planning a genetics unit and want every student to try a simulation.

You'd start by logging into your chosen platform—Labster is a favorite for many teachers because it covers biology, chemistry, and even health sciences.

There, you pick from dozens of ready-made labs; let's say you select a genetics simulation that covers Punnett squares and dominant/recessive traits. Assigning the lab is as simple as sharing a class code or link, which works on laptops, tablets, or even older desktops. Students log in with their credentials (no need to include full names if privacy is a concern) and enter the virtual experiment space. **The platform will guide them through an interactive scenario, prompting them to make predictions, adjust variables, and see instant results.**

Monitoring Progress Takes Only A Few Clicks

As students run the simulation, you can watch their choices in real time, who's adjusting gene sequences, who's skipping the instructions, and who's really experimenting instead of just guessing. There's even a space for students to record observations directly within the platform, such as notes on what happened when they changed a gene, screenshots of unexpected results, or brief reflections on what surprised them. For students who need more scaffolding, you can offer checkpoints, such as "Pause here and write what you think will happen if we double the number of dominant alleles," so everyone moves at his/her own pace but doesn't get lost along the way.

The real magic happens when AI steps in to personalize the experience. Unlike old-school simulations that play out the same for everyone, AI-driven labs shift based on student decisions. In an economics simulation, for instance, students might run a digital lemonade stand or manage a city's budget. If one group spends all their money on advertising but ignores product quality, the AI tweaks outcomes: maybe they get a sales spike followed by complaints or bad reviews.

Meanwhile, another student who budgets carefully and tests new recipes might see steady growth. The platform provides individualized feedback— *"Try varying prices," "Notice what happens when demand drops"*—so each learner's path is unique. In social studies, virtual field trips adapt to the students' interests; if someone asks about local history during a digital museum tour, the AI offers extra artifacts and deeper stories tied directly to their questions.

Assessment and reflection are built in. After finishing a simulation, students are prompted to connect their digital experience with classroom goals. For science labs, you might use reflection questions like: *"What variables did you change? How did those changes affect your results? What would you do differently next time?"* Students can submit short written or spoken responses within the platform—great for those who find writing tough—and teachers can respond directly with comments or follow-up questions.

For bigger projects, you might use a rubric that scores **not just accuracy but also curiosity and critical thinking**: *"Did the student make predictions? Did they adjust their* *strategy when things didn't work? Did they explain why outcomes changed?"* These rubrics help students focus on **learning as a process**, not just ticking off correct answers.

To make all this stick, consider using a **shared reflection template**. After each simulation, students jot down **three things they learned, one thing they found confusing, and one question they still have.** This not only builds metacognitive skills but also creates a **running record of growth over time**—a goldmine for parent conferences or IEP meetings. For teachers with limited prep time or large class loads, these platforms often provide ready-to-use templates and auto-graded reflections to lighten your load while keeping feedback personal.

AI-powered simulations invite all learners to experiment boldly and safely. When every student—whether reading below grade level, learning English for the first time, or working ahead—gets to test their ideas and see immediate feedback in action, real understanding follows. In my classrooms, I've watched students who barely spoke in group work sessions come alive while narrating their choices during a virtual lab. It's not about tricking kids into working harder; it's about making learning irresistible—messy, meaningful, and theirs.

Using AI Chatbots and Virtual Tutors to Support Struggling Students If you've ever wished for help answering student questions late at night or guiding a child through tricky math while you're working with small groups, **AI chatbots and virtual tutors can be invaluable.** These tools act as patient, always-available digital assistants, providing instant support and **allowing students to seek help whenever needed, without having to wait or fear repeating themselves.** AI chatbots, such as those found in **Duolingo or ScribeSense,** add a helpful technological layer to classrooms and homes by responding in real-time, repeating directions without frustration, and explaining concepts until students understand. **For those hesitant to ask questions in class, a chatbot offers a stress-free way to practice, review, and persevere through challenges.**

Using AI chatbots doesn't replace a caring adult; the best results occur when they're seamlessly integrated into classroom routines and homework support, as sidekicks, not substitutes. Start by introducing the chatbot as a resource. Set up "chatbot office hours" after school so students know when to log in for extra help. For those who finish early or get stuck while they're busy, the chatbot provides timely assistance. Some teachers have students keep a reflection log to track their interactions, which encourages metacognition and helps instructors identify trends, such as who needs repeated help with specific skills or topics.

A significant advantage of AI chatbots is their ability to personalize interactions. For students with learning differences—such as reading struggles, language barriers, attention issues, or anxiety—these digital helpers can easily adjust their language and pace.

With ELL (English language learners), for example, prompt the chatbot to use simple sentences or add home-language support. Many platforms also allow for toggling

"simple English" or integrating pictures for clarity. For students who find reading difficult or get overwhelmed by lengthy explanations, set the chatbot to break down information: **"Give one step at a time," "Use bullet points," or "Restate questions in simpler words."**

To support those who need extra processing time, consider slowing down the chat speed or allowing them as much time as necessary before responding.

Customizing chatbot prompts is simple. Rather than asking students to type open-ended questions, provide sentence starters: ***"Can you explain ___ differently?" "What does ___ mean in math?" or "I got this answer—can you show me where I went wrong?"***

For students with social anxiety or processing differences, structured starters make interacting with the chatbot easier. Special education teams can help create prompt templates tailored to specific accommodations, such as additional modeling or extra practice on foundational skills.

No tool works flawlessly at first. It's vital to monitor students' usage and refine your approach. Spend a few minutes each week looking at the chatbot's question log. If several students are asking about the same topic, consider revisiting that concept.

GATHER FEEDBACK WITH QUICK SURVEYS:

Check in with students who use the chatbot frequently or get sidetracked to see what's working and what isn't. Gather feedback with quick surveys: ***"Did this tool help your confidence?" "Was the information clear?" "What would you improve?"*** Use this input, paired with your observations, to **adjust classroom routines.** If students rely too heavily on the chatbot for answers, **encourage reflection** by asking them to write down their thoughts first or **explain their process** afterwards.

You may notice functional patterns, such as quiet students asking insightful questions at night, or others using the chatbot for quick spelling checks. These trends highlight where students feel comfortable exploring and where human interaction is still critical. When concerned about overuse or misunderstanding, set limits: establish session maximums, require teacher review for specific topics, or prioritize peer collaboration before turning to technology.

Ultimately, **AI chatbots and virtual tutors are the most effective when used as part of a broader toolkit, connecting teacher ability with student independence.** You remain the central guide and supporter; the digital assistant helps fill in gaps when your attention is stretched.

For parents, these tools provide reassurance that help is always accessible without replacing genuine relationships. **For educators and administrators,** usage data and student feedback enable targeted, actionable improvements, allowing instruction to be adjusted and growth to be celebrated, even when it's quiet and behind the scenes.

Student-Led AI Projects—Templates for Inquiry & Passion-Based Learning. There's something magical that happens when students get to run with their ideas. When you put real-world problem solving, creative thinking, and a **dash of AI into their hands,** you'll see energy and pride that no worksheet could ever match.

Student-driven AI projects aren't just about memorizing how technology works; they're about letting kids wrestle with questions that matter to them, design solutions, and take charge of showing what they know. **It's inquiry in action,** and it works for the full range of learners, from those who might need a scaffold to those who are itching for a challenge. **Giving kids ownership over an AI project** gives them a reason to care, to collaborate, and to surprise you (and themselves) with what's possible.

A great idea to kick off this kind of learning is with a **class "AI Invention Fair."** Picture a science fair, but instead of baking soda volcanoes, students dream up ways AI could help in their world, whether that's an app that supports classmates who speak different languages, a tool to recommend books based on mood, or a system to match volunteers with community needs. You'll see students brainstorming problems from their everyday lives, such as struggles with homework, organization, or even school lunch choices, and **thinking up tech-powered solutions.** Some will want to build; others may prefer to sketch or write. The real key is giving them space and structure to explore.

Templates help everyone get started, regardless of their level of comfort. **For younger students,** you might offer a simple outline: "What problem does your invention solve? How would the AI part work? Who would use it?"

For older students or those ready to go deeper, try a project journal: track your idea, research similar tools, test prototypes (even if it's just a paper model), and record what you learn along the way. In a language arts class, the project might be an AI-generated poetry anthology.

Each student could use an **AI writing tool** to create poems on themes they care about—such as friendship, change, or identity- then create the best ones into a digital book. Include space for students to reflect on which lines sounded "human" and which felt "machine-made."

For technology or ELA classes, "Design your own chatbot" projects are a hit: students outline what their bot should do (tutor math facts, answer school FAQs, offer daily affirmations), write sample scripts, then use basic AI tools or flowcharts to map conversations.

You don't need to direct every step; Your role is more that of a facilitator and coach. Begin with a brainstorming session where students list problems that bother or fascinate them. Then prompt deeper inquiry: **"Why does this problem exist?" "Who else has tried to solve it?" "What can AI do here, and what can't it do?" Encourage students to set milestones,** researching background info, drafting a plan, testing their tool or sample content, and preparing to share. Project journals are influential here.

Students document what tools they considered ("I tried three different AI writers before picking one"), record learning moments **("My first chatbot kept getting stuck on yes/no questions"),** and reflect on how their thinking changed. Peer review checklists keep collaboration honest and supportive: **"Did my partner's project address the problem? Was the AI part clear? What feedback could help them improve?"**

Once projects are ready, authentic sharing is everything. Students light up when someone beyond the classroom walls sees their work. You might organize a virtual exhibition. Parents and community members can log in to view live demos or posters that explain each student's project.

For those who freeze on stage, consider offering **digital galleries or Flipgrid-style video tours,** where students can explain their work at their own pace. Publishing an anthology online or sharing prototypes with local organizations makes the work matter even more. **Don't forget celebration**—certificates for **"Most Creative Use of AI," "Best Community Impact," or "Kindest Peer Reviewer"** remind everyone that effort and growth count as much as polished results.

Student-Led Projects Plant Seeds for Lifelong Curiosity and Problem-Solving These student-led projects don't just check a box for technology integration; they plant seeds for lifelong curiosity and problem-solving. Students learn it's okay to experiment, fail, ask for feedback, and try again—**skills that matter far beyond coding or AI prompts.** Your job isn't to have all the answers; it's to make space for questions and encourage every voice in the room.

As we conclude this chapter on creative engagement, remember: AI can do more than enhance a lesson; it can transform students into inventors and thinkers who see possibilities everywhere. **When you give your students the tools and trust, you open doors they'll keep walking through long after your class and the school year ends.** Students always remember impactful teachers.

Fostering Critical Thinking and Higher-Order Skills with AI

Chapter Seven

CHAPTER 7:

FOSTERING CRITICAL THINKING AND HIGHER-ORDER SKILLS WITH AI

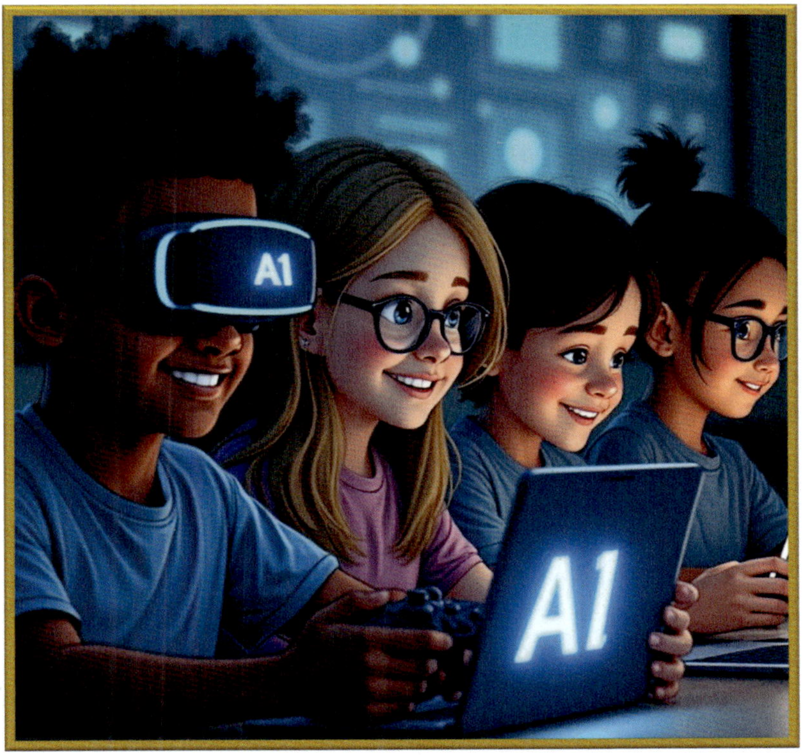

- ✓ **OBJECTIVE:** Students will analyze and debate a real-world topic using AI-supported research tools and structured dialogue.
- ✓ **VAKT ALIGNMENT:** Visual: Research maps; Auditory: Debate recordings; Kinesthetic: Live discussions; Tactual: Graphic organizers.
- ✓ **BLOOM'S LEVEL:** Analysis, Evaluation
- ✓ **INCLUSIVE SUPPORTS:** Promotes deep thinking in all learners through visual and oral critical thinking structures.

PHASE II: TEACH

Chapter 7. Fostering Critical Thinking and Higher-Order Skills with AI

STRATEGY 6. STRENGTHEN CRITICAL THINKING, NOT JUST ROTE ANSWERS.

DESIGNING AI-POWERED SOCRATIC SEMINARS AND DEBATES

Every teacher, parent, or administrator has wished at least once for a classroom where students engage in **deep conversations**, grapple with **big ideas**, and support every claim with solid evidence. But in reality? Sometimes you get blank stares when you ask, ***"What do you think?"*** or the same three kids jump in while others stay silent.

Here comes the twist: **Artificial Intelligence (AI)** can help stir curiosity, even in the quietest corner. Imagine you're preparing for a **Socratic** seminar on a classic novel or a heated debate about a newsworthy event. Instead of spending hours brainstorming questions, AI can craft provocative, open-ended prompts that spark genuine inquiry and thought.

CRITICAL THINKING TEACHING EXAMPLES WITH AI:

Say you're teaching *Animal Farm*. You feed the AI a quick prompt: **"Generate three essential questions for a 9th-grade Socratic seminar on power and corruption in** ***Animal Farm*****."** In seconds, you get gems like, ***"How does power corrupt individuals and groups in 'Animal Farm'?"*** or ***"Can a society exist without inequality?"***

These are not just Yes-or-No questions; **they invite conversation,** evidence, and disagreement. You can do the same for science, history, or even art.

AI doesn't just stop at classic literature; **it pulls fresh prompts from current events,** too. Plug into an AI news aggregator and ask it to find debate-worthy issues from this week's headlines.

Suddenly, you have ***"Should social media companies be responsible for fact-checking users?"*** or ***"Does artificial intelligence threaten creative jobs?"*** Students get to grapple with real-world controversies that feel urgent and alive.

DIFFERENTIATION IS KEY.

Not every student enters a seminar ready to roll. AI can help you assign roles—skeptic, clarifier, devil's advocate—based on student profiles or even at random to shake things up.

For students who need extra scaffolding (think ELL or SPED), you can ask AI for tiered question stems. Beginners might get, *"Why do you think...?"* or *"Can you explain that in your own words?"* while others work with, *"What evidence supports your view?"* or *"How could someone disagree with this point?"*

These tweaks transform passive listeners into active participants, providing everyone with a way in.

During the actual seminar or debate, facts matter. Let's say a student quotes a statistic about the environment or claims a character's actions have no parallel in history. Instead of relying solely on memory (or risking a heated "that's not true!" standoff), bring in an AI-powered fact-checker, like Perplexity or Bing Chat.

Students or teachers can toss a claim into the tool and, within moments, get credible sources to confirm or challenge what's just been said. It's like having a librarian and research assistant sitting right there at your elbow. This isn't about catching kids in mistakes; it's about modeling how to check your sources and build arguments on solid ground.

After all that lively discussion, **reflection matters** as much as the debate itself. AI can generate personalized reflection prompts: *"What changed your mind today and why?" "Who made you see something differently?" or "Which piece of evidence was most persuasive?"*

These questions help students dig into their thinking, not just **what** they said, but **how** they arrived at their views and **what they'd do differently next time.**

Teachers and students can use **self-rating rubrics** created by AI to assess the clarity of argument, the use of evidence, and even the willingness to consider opposing views.

It moves self-assessment **beyond** *"I did okay"* to real insight about strengths and next steps.

Interactive Element: Socratic Seminar Toolkit

Quick Start Checklist for Your Next Seminar or Debate:

1. Use an AI tool to generate three essential questions tied to your text or topic.
2. Ask AI for differentiated role cards and tiered question stems; print them out as needed.
3. Assign roles (skeptic, clarifier, devil's advocate) to mix up participation.
4. Set up an AI fact-checking tab for live debates.
5. After the discussion, distribute an AI-generated reflection sheet that asks students to analyze how their thinking has shifted.
6. Review self-rating rubrics as a group—celebrate growth and set goals for next time.

The magic isn't in the technology itself; It's in **how you use these digital tools** to unlock curiosity, encourage respectful disagreement, and help every student see themselves as a thinker worth listening to.

Even those who begin quietly or uncertainly can be significantly impacted by a well-timed, pertinent question. **And with AI lending a hand, you're free to focus on guiding students deeper, one bold question at a time.**

Using AI for Real-World Problem Scenarios and Project-Based Learning Picture this: your classroom hums with energy as students huddle around laptops, sketching blueprints and researching water filtration systems for a fictional town struggling with dirty drinking water.

No more blank stares o*r "When will we ever use this?"* Instead, students roll up their sleeves, eager to solve a problem that feels urgent, even if it's simulated. This is the magic that happens when you invite AI to help you build real-world scenarios for project-based learning.

You don't have to invent these problems from scratch. **AI platforms generate authentic, complex cases** that challenge students' thinking and bring textbook concepts to life.

For example, you can prompt an AI tool to *"design a hands-on environmental science project about global water access for seventh graders."*

In response, you'll receive a scenario about a community with limited resources and a challenge to engineer a sustainable water filter using everyday materials.

The scenario adheres to science standards and is age-appropriate, saving you time on adaptation.

The planning part can feel overwhelming when managing projects that cater to different student abilities and interests. That's where AI steps in as a project manager, not just a resource bank.

AI Breaks Big Task Into Phases

It can create **timelines with milestones and checkpoints**—think of it as a **GPS** for the entire project. The AI will suggest breaking the **big task into phases**: research, brainstorming, prototyping, testing, and presenting. **Each phase features built-in reminders and suggested deadlines,** which you can adjust to suit your pace or classroom schedule. If your group is new to inquiry projects, AI even generates step-by-step research guides.

For instance, it might suggest students start by identifying what clean water means, then investigate existing filtration methods, and finally collect data on different designs. Each research guide is **personalized**—some students may receive vocabulary scaffolds or sentence stems, while others may see advanced questions or links to recent studies.

Grouping students for collaboration is never simple, especially when you want everyone to feel included and challenged. Here's where AI's ability to analyze learner profiles is beneficial. It can recommend **groupings** based on strengths, interests, **Individualized Education Programs (IEPs)**, or language proficiency, thereby ensuring diversity within each team. Need someone to do research?

The platform matches the role with a student who excels at gathering information and looking for a presenter. AI nudges forward more vocal students or those ready to stretch themselves.

Even shy or less confident kids can get purposeful roles, such as a data analyst or a visual designer. Throughout the project, AI can run sentiment analysis on group chat logs or digital journals. If one team member feels left out or frustrated, the system flags it for you, allowing you to check in before minor problems escalate into major ones.

When it's time for students to show what they've created, your options multiply with AI on hand. Instead of traditional poster boards or essays, students can build digital portfolios that collect every phase of their work—notes, designs, photos of prototypes, and reflection videos.

Presentations Take on New Life With AI.

The AI platform organizes all. Students upload slides or videos to an AI tool that gives instant feedback based on their criteria (clarity, persuasiveness, use of evidence). The input goes beyond grammar—it might say, ***"Your solution explained costs well but could use more detail on maintenance,"*** or ***"Great use of statistics! Next time, try connecting them to personal stories."***

This kind of response helps all students grow, from your quietest ELL learner to your most outspoken gifted student.

Sharing Solutions Globally With AI

Sharing solutions with the broader world **gives real purpose to these projects.** AI can help you publish student work on a class website—editing summaries for clarity and polish—or even generate press releases for community newsletters.

If you want outside input, consider using AI to facilitate community feedback forms or connect with local experts who are willing to provide digital comments on student solutions.

This public audience motivates students to care about quality and impact; suddenly, their water filter matters because someone outside the classroom is listening.

Project-based learning powered by AI

Project-based learning empowers students to see themselves as problem solvers. Instead of memorizing facts for a test, they develop persistence, teamwork, and creativity—skills that matter far beyond graduation day.

And all the while, you save time on setup and tracking while gaining more energy to mentor and cheer them on as they figure out real answers to real problems.

Embedding AI-Powered Reflection and Metacognition Checkpoints If your students forget lessons quickly, you're not alone. Sometimes all it takes is a brief pause for students to reflect, asking themselves, "What did I do wrong?"

Self-Checks

"Where did I get stuck? Did I try something new?" These self-checks build critical thinking and self-awareness.

AI can help weave these reflective moments into any lesson, regardless of the subject or grade level. Instead of collecting stacks of paper exit tickets, you might use an AI tool to prompt students with questions tailored to your lesson, such as "What strategy helped you solve today's math problem?" or "What was the trickiest part of the science lab and how did you deal with it?"

The AI adjusts these prompts to align with your learning goals and students' language capabilities, nudging each student toward deeper thinking.

When students respond, AI rapidly shows patterns that you might overlook—building word clouds of commonly used phrases, such as "guessing," "drawing a picture," or "asking a friend."

This snapshot helps you see how students approach problems and identify where difficulties tend to cluster. Over time, the system tracks changes in thinking, such as an increase in answers mentioning "explaining my answer" as students become more confident.

You instantly see which strategies are taking hold and which need reinforcement, sparing you hours sorting through physical notes.

AL's Impact

The impact of AI is even more substantial with personalized feedback. Instead of writing short comments like "Good job," AI reviews each reflection and generates growth-focused observations: "You tried a new approach when stuck—great perseverance!" or "You asked questions when confused—keep advocating for yourself!"

This feedback values adaptability, problem-solving, and self-regulation, not just correct answers. You can fine-tune comments as needed so that students feel recognized for their thinking, not just what they produce.

Reflection can also be communal. AI suggests group discussion prompts for digital journals or chats, such as "What did your group learn about teamwork?" or "How did your team resolve disagreements today?" If a conversation stalls, the AI can propose follow-up questions such as, "Did anyone change their mind after hearing another idea?"

These prompts help students view themselves as part of a thinking community, not just working alone. For students who struggle to express themselves—mainly English learners or those with learning differences—AI-generated sentence starters (**"I noticed that...", "Next time, I would try..."**) lower the bar for participation.

Some teachers worry that technology will diminish authenticity or control. But you decide which prompts to use, how often to reflect, and what follow-up looks like. AI supplements your ability, making it easier to capture critical metacognitive moments. Reflection is challenging for most, so a tool that gently prompts students can help build habits that last a lifetime.

Peer feedback is another strength of AI. Students working in pairs or groups can share responses in a digital space where AI prompts them with questions, such as, **"What else could we have tried?"** or **"How did we manage mistakes?"** You can quickly review exchanges, stepping in when needed, but students increasingly learn to offer detailed, supportive feedback themselves.

AI-generated charts assist visual learners by showing class trends, such as the number of students mentioning perseverance or which group adjusts plans efficiently. These visuals celebrate collective progress, highlight areas for class improvement, and identify students who require additional support.

Teachers can also use these AI-collected insights for their professional development. Reflect on which AI prompts lead to deeper discussions and adjust reflection points as needed, after group tasks sometimes, and after solo work at other times.

Flexibility allows for reflection to fit naturally into your class schedule without demanding elaborate rituals.

Parents and caregivers also benefit: AI-generated summaries can serve as conversation starters at home, such as "***Ask your child about a strategy they used to solve a problem this week.***" This fosters a partnership with families, promoting metacognition beyond the school setting.

Ultimately, building regular, personalized reflection into the routine makes critical thinking and self-awareness an everyday practice. AI isn't magic; it's an amplifier, making thinking visible, attainable, and worth celebrating each day.

Assessing Critical Thinking Growth with AI Analytics & Teacher Guides

Tracking the growth of critical thinking is challenging—it appears in moments when students connect ideas or question sources, but measuring this reliably across all students is difficult.

That's where AI analytics play a key role, serving as a behind-the-scenes coach. Imagine a dashboard gathering evidence from essays, digital discussions, and project submissions, analyzing language for reasoning, synthesis, and creativity.

Clear patterns emerge, maybe students cite facts but struggle to synthesize information, or perhaps they're improving at using evidence but need help drawing original conclusions. This provides a real-time pulse check, allowing you to identify growth and gaps efficiently, without being overwhelmed by paperwork.

AI platforms now offer visual "critical thinking dashboards."

These tools provide bar graphs and heat maps that illustrate, for example, the percentage of students who support claims with evidence or highlight weak reasoning in specific topics.

Instead of having to guess who needs help, you receive targeted insights, like John consistently building strong arguments but lacking creativity, or Aisha being original yet needing more evidence. These tools enhance (not replace) your judgment, clarifying student progress and saving you time. Dashboards transform abstract skills into concrete, trackable progress for all learners, including ELLs and students with IEPs, ensuring you know who needs extra support or new challenges.

AI can also help design **effective rubrics** that capture **critical thinking**. A robust rubric breaks down skills like reasoning, use of evidence, synthesis, and creative problem-solving. Criteria might include **"connects multiple sources," "challenges assumptions,"** or **"proposes alternative solutions."**

AI Rubric for Critical Thinking

AI can suggest rubric criteria and age-appropriate language, streamlining the process for different groups. Once your rubric is set, you can upload it to your platform, enabling AI to score student work according to these indicators.

After assignments, you might get summaries like: **"70% of students showed strong reasoning; 40% responded with multiple perspectives; creative problem-solving increased by 15% this month."** These aren't just numbers—they highlight where instruction is working and where more support is needed.

AI systems need to be calibrated, and AI must understand what **"good"** looks like in your classroom. Regularly review a sample of AI-scored work alongside your ratings to ensure alignment.

You may find the AI favors length over depth, or it overlooks nuances in non-traditional answers. Periodic check-ins let you refine rubrics and retrain the system using real examples from your students. This collaborative approach keeps accuracy and fairness while honoring your professional ability.

Communicating progress turns data into value.

Templates for student reports or parent conferences let you share not only grades but also stories of growth in analysis and thinking.

Reports might say, *"Sofia now routinely questions sources; her last project included three unique viewpoints."* At student-led conferences, AI-generated visuals can help learners explain their progress: *"Here's how my ability to connect ideas improved this semester."*

This approach ensures **every learner is recognized for specific strengths and next steps,** not just high performers. **Parents see concrete examples; administrators see clear evidence of higher-order learning.**

Keep Communication Simple and Jargon-Free.

Focus on student growth, what they can achieve now, and what's next. Provide specific praise **("You blended ideas from history and science in your project!")** and actionable feedback **("Next time, use a wider variety of evidence to back up your claims").** Suggest easy ways for families to support development at home, like asking about challenging learning moments in the week.

Visual Element: Sample Critical Thinking Dashboard

Example dashboard for a sixth-grade class:

- Reasoning: 68% meeting or exceeding expectations
- Evidence Use: 54% showing growth.
- Synthesis: 33% need targeted support
- Creativity: 71% above baseline

This gives you an immediate sense of where to target next week's lessons.

Combining AI Analytics with Teacher-Designed Rubrics

Overall, combining AI analytics with teacher-designed rubrics doesn't just make critical thinking measurable; it **makes it visible and actionable for all.** Rather than relying on memory, every student's progress is tracked and built upon.

These insights help you adapt instruction, making continual improvement a part of everyday teaching, seamlessly leading into refining practice with **Continuous Quality Improvement (CQI)** in the next chapter.

"Artificial intelligence is the new electricity."
— Andrew Ng, AI Researcher.

INTEGRATING SEL (SOCIAL-EMOTIONAL LEARNING) & WELLNESS THROUGH AI

CHAPTER EIGHT

CHAPTER 8:

Integrating SEL (Social-Emotional Learning) & Wellness through AI

- ✓ **OBJECTIVE:** Students will reflect on their emotions and set wellness goals using AI journaling or SEL coaching platforms.
- ✓ **VAKT ALIGNMENT:** Visual: Mood trackers; Auditory: AI-guided mindfulness; Kinesthetic: Yoga/movement check-ins; Tactual: Printed journaling sheets.
- ✓ **BLOOM'S LEVEL:** Comprehension, Application
- ✓ **INCLUSIVE SUPPORTS:** Empowers SEL, ESOL, and SPED students to express emotions safely.

AI Tools for Monitoring Student Well-Being and Emotional Check-Ins

Phase II: TEACH

Chapter 8. Integrating SEL & Wellness through AI

Strategy 7. Integrate Social-Emotional Learning and Mental Wellness.

AI Tools & Software:

- Peekapak
  - **Description:** SEL curriculum infused with AI personalization options.
  - **Link:** https://www.peekapak.com/
- Along
  - **Description:** AI check-ins and reflections to promote student well-being.
  - **Link:** https://www.along.org/
- MindSpark Learning
  - **Description:** Professional development for AI and SEL fusion.
  - **Link:** https://mindspark.org/

Imagine your typical classroom: The buzz of chatter, a water bottle spills, and a couple of kids already carrying emotional burdens before classes even start.

While academics are important, what matters just as much is what's going on beneath the surface. AI-driven emotional check-ins and wellness trackers are changing how teachers support students' less visible needs.

AI platforms like **Along and Mood Meter** help students, especially those who tend to keep things inside, express how they're feeling through quick, regular check-ins.

On the Mood Meter: For example, students tap a spot on a color-coded grid (red: angry, blue: sad, yellow: happy, green: calm) to identify their mood, then can specify it more precisely and jot down reasons.

This process is private, quick, and stigma-free, benefiting both teens and adults. AI "pulse" surveys provide another way to integrate check-ins into your daily routine.

These mini-surveys—run on **Chromebooks, iPads, or shared devices**—ask simple questions, such as **"How are you today?"** or **"What's one word for your mood?"** Anonymous options allow students to be more honest, which is crucial for those concerned about privacy or their peers' reactions. Data is sent only to the teacher's secure dashboard, making it easy to keep responses confidential.

These check-ins fit seamlessly into routines, scheduled during transitions or at times when students are settling down, such as after recess, during advisory, or as a calm-down at the end of the day. Frequency is flexible (daily or weekly), but consistency is crucial, so students can trust that the space is always available. Participation should always be voluntary; pressure to share undermines the integrity of the process.

Privacy is critical: stick with platforms that are FERPA-compliant and avoid unnecessary data collection (1). Use first names or initials where possible and remind students and families that only immediate safety concerns override confidentiality.

AI turns all this data into actionable insight without overwhelming you. Dashboards produce weekly summaries, allowing you to spot patterns, such as whether more students experience feelings of depression in winter or become anxious before significant assignments. Real-time alerts flag when someone's mood suddenly drops or persistent issues surface, allowing you to identify students who may need extra support before the situation escalates.

What's next when **AI data highlights a student in distress**? AI is a starting point—your intuition still matters. If a student is repeatedly flagged for negative emotions, reach out with a discreet conversation: "I noticed you've had some hard days. Want to talk?" You can use simple scripts: "Is something at school or home making things tough?" or "What's something that could help today?" If the student isn't ready, your concern alone shows you care.

If a **trend appears across the class**—say, widespread anxiety before a project—address it together: "It looks like many of us are worried about the assignment. Let's talk strategies and support each other."

For complex or ongoing issues, AI-generated summaries can help streamline referrals to counselors by providing clear, organized overviews: dates, emotion types, and student comments. This saves everyone time and ensures no details get missed (3).

Interactive Element: SEL Reflection Prompt

Try this five-minute exercise with your class, or just yourself:

"What color matches your feelings right now? Why?"
Have everyone jot down their answer (or do it yourself). See if mood patterns or themes appear over time—are certain days or assignments tougher? Use these insights to refine your routines, offer encouragement, or reinforce the message: every feeling is valid, and support is always close by.

AI isn't a cure-all, but it's a powerful tool for insight. It helps to surface emotions students may not voice, creating a routine where honesty is valued instead of hidden. Integrating well-being into your daily schedule strengthens relationships, gets students help when needed, and fosters a classroom culture built on trust. Everyone receives help from a community where emotions can be safely named and addressed.

Customizing SEL Interventions with AI—VAKT Lesson Templates

There's something powerful about watching a student's face light up because a lesson finally makes sense—not just in their head, but in their heart and body too. **That's the magic of using VAKT—Visual, Auditory, Kinesthetic, and Tactile**—approaches in social-emotional learning. Now, with AI, you don't have to scramble for ideas or spend your Sunday searching Pinterest for that elusive **"perfect empathy activity."** AI can generate entire SEL mini lessons, blending these modalities in ways that fit your unique students, no matter their age or ability.

Start by thinking about how your students connect (VAKT): some process big feelings through **drawing or visual stories,** others need to talk it out or move around. When you prompt an AI tool, you can ask for lesson blueprints that include a visual anchor, like an illustrated comic strip about sharing or managing frustration. **For auditory learners,** request an audio story or reflection script. AI can quickly whip up a mindfulness narration, perfect for a calming five-minute reset during homeroom or after lunch. **If movement is key** for your crew, ask the tool for **kinesthetic activities—a "calm-down yoga"** routine or even a role-play where students practice assertive communication while standing and moving. Tactile learners may receive help from creating emotion jars, stress balls, or journaling with textured materials. **AI makes it easy to tailor each activity, so you're not left guessing which approach will reach the student who constantly fidgets or the one who tunes out unless their hands are busy.**

Let's say you're preparing a **week of SEL moments focused on friendship and conflict resolution.** Open your favorite AI platform and prompt it for a lesson sequence: ***"Give me a visual scenario about two friends disagreeing, with comic panels or emotion thermometers."*** Then ask, ***"Write an audio reflection script that guides students to think about how they'd feel in each friend's shoes."*** For your **kinesthetic piece,** consider a movement-based routine—perhaps a walk-and-talk format where pairs move around the classroom, sharing times they've worked through a disagreement. If you want a tactile component, consider a hands-on craft, such as making apology cards or building a friendship web from a string. **AI generates these templates in seconds, providing the framework for you to personalize them for your class.**

Adapting these ideas is just as simple. Suppose you have a student who struggles with traditional meditation scripts but loves movement and music. Adjust your AI prompt: ***"Rewrite this mindfulness script as a standing stretch-and-breathe routine set to music."*** The tool will deliver easy-to-follow steps you can print or project. If you **teach ELL students or those with limited reading skills,** consider requesting picture support or audio tracks in their native language. For students with **sensory sensitivities,** consider generating scripts that utilize soft objects or textured materials during reflection time, rather than engaging in loud group activities.

Sequencing and scheduling also matter. AI helps you plan routines that build emotional muscles over time. You might ask for a Monday morning **"reset"**—a brief visualization exercise to set intentions for the week—or a Friday gratitude circle that combines writing (for those who prefer quiet) and group sharing (for your more vocal team members). These small moments, embedded regularly, shift the classroom culture and help students naturally internalize SEL skills.

Real classrooms bring these strategies to life in creative ways.
An example of a classroom activity for **second graders** could be to use an AI-generated visual **"emotion thermometer"** every morning. Kids moved clothespins with their names onto colors representing their moods—blue for tiredness, yellow for excitement, and green for calm. Instead of lengthy explanations, students could convey their feelings in seconds, and the teacher could identify who needed extra support that day without putting anyone on the spot.

In another case, a fifth-grade teacher could help alleviate students' test anxiety by listening to an AI-guided auditory visualization—**imagining themselves** in their *"Happy Calm Place"* before taking their math test. Students often report that this approach helped them focus better and feel less anxious.

Teachers find that these VAKT-based activities also support kids with learning differences. One student with ADHD thrived on tactile journaling—adding stickers or textured tape to mark big feelings—while her classmate with dyslexia preferred listening to audio reflections and drawing her responses instead of writing them out. **The key is flexibility: AI enables you to tailor every lesson so that no one gets left behind.**

If you want to put this into practice right away, try this prompt: ***"Create an SEL mini lesson on self-control with one visual activity, one guided audio script, one movement break, and one hands-on project."*** See what arrives and adjust it to fit your group's quirks and strengths. You'll be surprised at how easily SEL becomes part of your daily rhythm—and how much more engaged your students become when their needs are truly seen.

AI for Building Empathy—Digital Scenarios & Guided Role-Play

When it comes to empathy, nothing beats that spark of **"aha!"** when a student realizes what life is like for someone else—maybe a classmate, a neighbor, or even a character in a story. These are the moments that shape not just better students, but kinder people. AI can bring these lessons to life by generating digital scenarios that are anything but generic.

Imagine an interactive story where you get to **"walk around"** as a new kid who speaks a different language, or perhaps as someone who has just lost their pet, or as a student facing a tricky friendship problem. With just a few **clicks** and the right prompt, you can create age-appropriate stories where students make choices as characters from all sorts of backgrounds. Each decision alters the path of the story, illustrating how words and actions ripple out and impact those around them.

Empathy Isn't Automatic. One thing I've learned over the years is that **empathy isn't automatic.** Kids—and adults, too—need practice. AI helps by scripting out role-play exercises that are detailed and realistic, taking the pressure off teachers to come up with everything themselves. You can ask your AI tool for a set of character cards: maybe one card describes Maria, who feels left out at recess, while another outlines Juan, who wants to help but isn't sure how. Each card includes not just names but emotions, motivations, and even little challenges (like ***"English is not your first language" or "You're worried you'll get laughed at").*** You hand out these cards and read through the script together: ***"Maria wants to join the game. What could Juan do?"***

The script offers possible responses, leaving room for students to improvise—sometimes with hilarious, sometimes with moving results. If you want to delve deeper, try building a complete **conflict resolution circle** or a **restorative justice scenario. AI can generate scripts where students play roles**—maybe one is the person who was hurt, another is the person who caused harm, and others are witnesses or supporters. The script includes guiding questions: ***"How did you feel when this happened?" "What would help repair the hurt?"***

By following these steps, everyone has a chance to speak and listen in turn. The teacher's job is primarily one of guidance, not direction. This structure provides a safe way for even the shyest or most anxious students to participate.

Relevancy is key. It's no good if scenarios sound like they were written in another country or century. AI lets you tailor everything to your students' world. With a quick prompt, you can incorporate cultural traditions, perhaps have students plan a role-play around a family holiday or a community event that truly matters to them. **For ELL (English Language Learners) or students from immigrant families**, AI can weave in home languages or customs: ***"Write this scenario using both English and Spanish phrases,"*** or ***"Include a character who brings a traditional meal to class."*** Students light up when they see their own lives reflected in what they're learning. It also helps everyone else in the class step outside their bubble.

Reflect and Measure Growth

After these activities, it's essential to **reflect and measure growth,** not through tests, but through honest conversations and simple tools. I like to hand out journaling prompts right away: ***"How did it feel to be in someone else's shoes?" "What surprised you?" "What would you do differently if this happened in real life?"*** You can use digital forms or good old paper and pencil. For younger students or those who need added support, try using sentence starters such as ***"When I was playing [character], I felt..." or "I learned that..."*** Give them a few minutes to write or draw their response.

Peer feedback helps, too. Have students fill out short forms about their classmates' listening and perspective-taking skills: Did their partner listen without interrupting? Did they try to understand another point of view? Collect these forms and look for patterns over time—are students getting better at naming feelings? Are they more willing to apologize or compromise? This kind of assessment isn't about grades; it's about growth.

I've seen firsthand how these **empathy-building activities** change classroom dynamics. The quietest student might suddenly speak up with insight because the scenario matches their life—children who never thought twice about teasing come to realize how their actions sting. Even adults get something out of it—sometimes we need reminders to pause and see things from another angle, especially on tough days.

With AI as your co-planner, you don't have to do all the heavy lifting alone. You get fresh ideas, scripts you can tweak on the fly, and resources that help every student feel seen and heard. The magic comes when you step back and watch your class connect on a new level—not just as learners, but as real people trying to understand each other just a little bit better.

Addressing Mental Health Early—AI Alerts and Teacher Action Plans

Sometimes, you sense that something is **"off"** with a student—a sudden shift in attitude, silence, or troubling writing. AI tools can spot these patterns early, even when you're stretched thin. By analyzing language in journals, essays, or discussion posts, AI finds clusters of words that may indicate distress, hopelessness, or anxiety. It acts as an extra set of eyes, flagging moments when a student's written tone noticeably changes.

For instance, a rise in words like ***"tired," "worthless,"*** or ***"can't"*** could prompt an AI alert, inviting you to look more closely. AI doesn't diagnose; it simply surfaces potential concerns for your attention.

When you receive an **AI alert,** don't panic; it's an opportunity to pay more attention, not an emergency declaration. Review the flagged assignment calmly, looking for context and patterns instead of jumping to conclusions. Use a checklist: Has the student's tone changed drastically? Are themes of sadness, anxiety, or isolation recurring? Does the writing mention specific worries about friends, home, or school life? Patterns matter more than a single alarming entry—sometimes students vent dramatically without deeper risk, while others might hide distress behind humor.

Documentation is essential, both for your memory and to provide clarity if others need to be involved. Quickly jot down the date, type of assignment, summary of the AI alert, and your main observations; no need for detailed narratives, bullet points suffice. If extra support is needed, these notes help counselors gain a comprehensive understanding of the situation. Many schools use digital templates for tracking concerns; if not, a simple shared document works.

Trust Your Instincts But Follow Established Protocols.
When determining the next steps, trust your instincts but follow established protocols. For immediate safety concerns, follow your school's emergency procedures and never leave a student alone if harm is likely to occur. For less urgent but concerning trends, please reach out to your campus counselor with a summary drawn from your notes and the AI alert. Most counselors prefer concise emails like: ***"Several recent assignments from [student] mention hopelessness and isolation. The AI flagged these, too. Can we check in together?"*** This teamwork speeds up support and maintains clear communication.

Families are crucial partners, and sensitive communication is key. When contacting parents about mental health issues identified through AI, maintain a tone that is both factual and caring. For example, a phone call might be: ***"Hi, I've noticed some recent changes in [student]'s work—they're expressing more overwhelm than usual. Nothing urgent has happened, but I wanted to check in and let you know I'm here if you want to talk or connect with school support."*** For emails, be brief and warm: ***"Some trends in [student]'s assignments suggest they may be going through a tough time. I'm reaching out to see how things are at home and offer support if needed."***

Confidentiality is essential throughout. Only share details with those who need to know—students trust us when we protect their privacy. If you're unsure how much you can share, consult with your counselor or administrator.

Ultimately, tech alone can't solve mental health challenges. What matters is how we act on the information. Creating a classroom climate where asking for help is safe and regular is just as vital as responding to alerts.

Small routines help: start class with a positive affirmation ("Today I'll be kind to myself") or have students write a weekly gratitude note in a shared journal. AI can suggest prompts or reminders for coping exercises, such as group breathing exercises before tests or proactive teaching strategies to prevent stress from escalating.

Destigmatizing mental health means open and everyday conversation, never making it awkward or scary. Use language that normalizes struggle: ***"Everyone has tough days,"*** or ***"It's okay to ask for help."*** Post resources in your classroom or online, such as helplines, local counseling information, and calming apps. **Modeling openness** and resilience yourself—acknowledging occasional hard days in age-appropriate ways—shows students that reaching out is a sign of strength.

Overall, **AI offers more opportunities to identify minor issues before they escalate into larger problems.** It connects the dots across assignments and weeks, offering prompt insights while respecting student dignity. When combined with caring conversations and structured routines, technology is an added layer of classroom support—not a substitute for genuine human care, but a powerful ally. Remember, wellness impacts every aspect of learning. The following chapter will explore how global collaboration can further enrich your classroom.

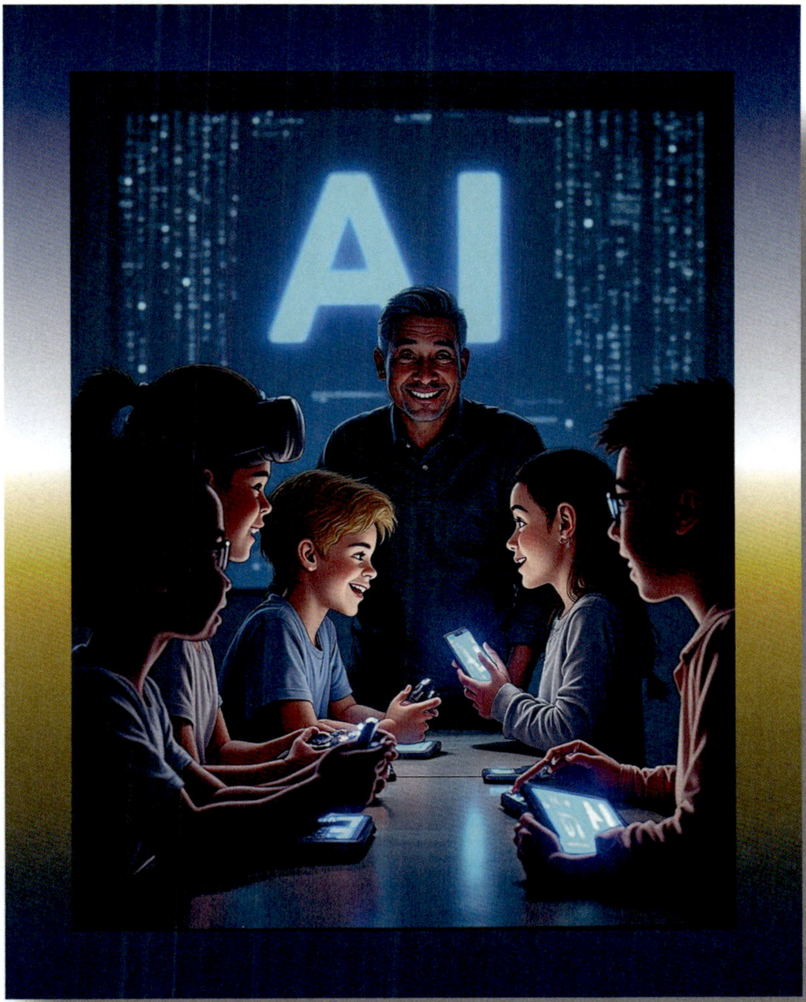

Global Collaboration– Connecting Classrooms with AI

Chapter Nine

CHAPTER 9:

GLOBAL COLLABORATION–
CONNECTING CLASSROOMS WITH AI

- ✓ **OBJECTIVE:** Students will collaborate with peers globally to co-create a digital project on a shared social or academic issue.
- ✓ **VAKT ALIGNMENT:** Visual: Interactive digital tools; Auditory: Live language exchanges; KINESTHETIC: Cultural simulations; Tactual: Tangible cultural artifacts.
- ✓ **BLOOM'S LEVEL:** Synthesis, Application
- ✓ **INCLUSIVE SUPPORTS:** Bridges equity gaps and increases global literacy for all learners.

AI Tools & Software & Resources For This Chapter on Global Collaboration

Phase II: TEACH

Chapter 9. Global Collaboration—Connecting Classrooms with AI

Strategy 8. Collaborate with Classrooms (and Experts) Around the Globe.

- Empatico
  - **Description:** Connects classrooms globally for empathy and cultural learning.
  - **Link:** https://empatico.org/
- PenPal Schools
  - **Description:** Global collaboration platform for project-based learning.
  - **Link:** https://www.penpalschools.com/

AI Translation Tools

AI translation tools, such as **Google Translate** and **Microsoft Translator**, provide real-time translation for chats, documents, and live video.

DeepL is another accurate tool, particularly for complex texts.

For video calls on **Zoom** or **Teams**, enable live captioning and choose your preferred language: in Zoom, click **"Live Transcript"** and select the target language; in Teams, select **"More Actions,"** then **"Turn on live captions"** with your language preference.

This allows even shy or language-learning students to keep up and participate.

Setting Up Virtual International Exchanges with AI Translation Tools

Imagine students crowding around a webcam, their faces lighting up as they interact with peers from around the world. Laughter and curiosity fill the air as they exchange greetings in different languages and share favorite snacks. In these moments, classroom walls disappear, giving everyone a sense of belonging to something far larger.

This connection isn't just uplifting, it's crucial for building global competence and empathy. In today's world, connections—**bridging cultures, sharing ideas, and communicating across languages**—are essential. **Virtual international exchanges,** powered by AI translation tools, provide students and teachers with a unique and personalized way to engage in global learning.

Why Engage with Classrooms Abroad With AI?

These exchanges bring lessons to life. When students communicate directly with peers in Japan or exchange holiday traditions with a class in Nigeria, they develop language skills, curiosity, compassion, and genuine motivation.

Language practice stops being rote; instead, students naturally ask questions, share experiences, and sustain conversations. These opportunities are especially valuable for ELLs, SPED students, and those who haven't traveled—exposing everyone to new perspectives and igniting excitement.

Students realize their words matter to someone far away, fueling motivation and broadening worldviews.

How Do You Set Up These Exchanges with Ease Using AI?

The key is to use educator-focused platforms—no travel funding required. Platforms like ePals (connecting K-12 students in over 99 countries), iEARN (with more than 150 collaborative projects), and PenPal Schools (encompassing students from over 150 countries) simplify the process.

They guide you through creating class profiles, managing privacy, and matching by grade, subject, or interest. Before reaching out, **craft a welcoming and concise introduction**, such as:

"Hello from Maryland! I teach fifth grade at a Title I school, and my students would love to exchange letters and projects with your class. We're eager to learn about your culture and share ours with you. Would you be interested in setting up a monthly video call or collaborating on a project? Looking forward to hearing from you!"

Keep it friendly, direct, and open to questions. Teachers everywhere may be just as eager (and a little nervous) as you are.

With a match secured, address language barriers directly—no extensive vocabulary is needed.

Shared Projects or Digital Presentations

For shared projects or digital presentations, AI translation enables the conversion of documents into multiple languages in seconds. For example, if both classes develop a slideshow about school lunches, students write captions in their native language, then use Google Docs' translation function for their peers.

This not only includes everyone but also fosters respect for all languages.

In practice, exchanges can start simply—with a "culture show-and-tell." Each class prepares a brief presentation about a routine, holiday, or favorite food. Students can utilize slides, short videos, and AI tools to translate captions or subtitles, ensuring that every class is accessible to all students.

For ongoing connections, introduce a weekly "question of the day" (e.g., "What does recess look like at your school?" or "What's the weather like where you are?"), using AI translation to exchange questions and responses smoothly.

Establish communication protocols and etiquette. Remind everyone that cultural norms vary—polite habits differ worldwide. Start interactions with greetings and gratitude, avoid confusing slang, and encourage curiosity over judgment. Clarify response times (mind the time zones!), preferred formats (written, video, voice), and privacy rules (safeguard contact details). Set an example of patience and good humor for inevitable translation hiccups—these are also learning moments.

Interactive Element: Global Exchange Reflection Prompt

After the first exchange, have students (and yourself) note three surprises from their partner class, one challenge (such as a tricky word or cultural difference), and one question they would like to ask next. This reflection helps everyone process the experience and look forward to future exchanges.

By combining AI translation with well-structured international exchanges, you allow every student to communicate globally, regardless of their language background or learning challenges. This is how classrooms evolve into centers for empathy, confidence, and meaningful, worldwide conversations.

Building a Safe Online Collaboration Culture—AI Moderation and Best Practices

When students and teachers connect globally, new responsibilities emerge. Beyond lesson planning and group work, educators must guide everyone through a digital space that combines cultures, languages, and expectations, much like hosting a worldwide online event where everyone expects respect and safety. The internet, however, poses risks: cyberbullying, oversharing, and cultural missteps can escalate before they're noticed. Therefore, building a safe, respectful online culture is as essential as the subject matter itself.

Teachers serve as digital conductors, setting the tone and expectations from the outset. Safety isn't a single lesson; it must be woven into every phase of global collaboration. Clear interaction guidelines are crucial: always use respectful language, seek consent before sharing photos or personal details, and remember that humor or emojis may not translate well across cultures. Everyone—students, teachers, and parents—needs to understand their role. Students act as digital citizens, treating others with respect and reporting any suspicious activity or concerns. Teachers are not only mentors and first responders but models of fairness and transparency.

Establishing collaboration rules should be a team effort, not just a teacher mandate. Begin with a "netiquette" contract, signed by all, expressing guidelines like using real first names, providing constructive feedback ("I like how you explained this…" instead of "That's wrong"), and never sharing screenshots or recordings without permission.

For rule-breaking, follow a stepwise protocol: first, a private message; if issues persist, notify parents or your partner teacher, explaining the steps taken. The aim is accountability and growth, not punishment, fostering an atmosphere where admitting mistakes feels safe and secure.

Despite strict rules, some problems will persist. This is where continual monitoring and feedback are key. Monthly reviews of AI-generated reports can help highlight participation imbalances, flagged chats, and recurring issues, such as misunderstandings over slang or instances where one person dominates conversations. Use these insights to regroup, revisit guidelines, or encourage fewer vocal students. Anonymous surveys help too: ask students how safe they feel, if they've encountered disrespect, and for suggestions to improve inclusivity.

Culture is dynamic and shifts with each group or project. Adjust approaches regularly. If reports show increased conflict near deadlines, consider adding extra check-ins. If students seem unclear on privacy, revisit those rules. Encourage students to reflect on their online behavior: do they pause before posting or check the tone of their comments? This isn't about micromanaging speech but about fostering digital empathy—recognizing the person behind each screen.

Teachers should also demonstrate vulnerability by admitting when they are unsure about how to handle ambiguous issues. If an AI tool flags a debatable comment—a joke taken wrong or a cultural disagreement—bring it to the group for discussion. This fosters trust and empowers everyone to take responsibility for maintaining a safe online environment.

Reflection Section: Quick Digital Citizenship Pulse Check

At the end of each month, ask students to rate their sense of online safety (1-5) and share one thing that made them feel included, as well as one thing they wish would change. This simple reflection reveals hidden concerns and provides valuable, prompt feedback to strengthen the culture of collaboration.

In summary, safe global collaboration stems from established routines, candid conversations, and tools that protect while fostering connection. Proactive use of AI moderation and clear agreements fosters an environment where everyone's voice is respected, and all students grow as digital neighbors.

Next, we'll explore using AI analytics for tracking and celebrating collective progress, reinforcing the idea that learning together should move us all forward.

"Where AI sees data, the human heart sees meaning."
— Anonymous modern proverb (AI ethics circles)

"AI offers more opportunities to identify minor issues before they escalate into larger problems."

Data-Driven Assessment
& Growth Tracking with AI

CHAPTER TEN

CHAPTER 10:
DATA-DRIVEN ASSESSMENT & GROWTH TRACKING WITH AI

- ✓ **OBJECTIVE:** Teachers will utilize AI analytics to interpret assessment data and plan responsive instructional adjustments for at least one subgroup.
- ✓ **VAKT ALIGNMENT:** Visual: Dashboards; Auditory: Coaching videos; Kinesthetic: Data sorting tasks; Tactual: Printed analysis reports.
- ✓ **BLOOM'S LEVEL:** Analysis, Evaluation
- ✓ **INCLUSIVE SUPPORTS:** Improves targeted instruction and RTI for diverse populations.

Using AI Dashboards for Individual and Group Progress Monitoring

Teaching a room of thirty students can be a challenge—you know some are excelling while others are quietly struggling, but it's hard to keep track. This is where AI dashboards become invaluable. Think of them as your assistant, keeping track of all the minor data points—quizzes, homework, class participation, exit tickets—so you can focus on instruction.

AI Tools & Software for This Chapter on Data-Driven Assessment

Phase III: ASSESS

Chapter 10. Data-Driven Assessment & Growth Tracking with AI

Strategy 9. Utilize AI Assessment to Evaluate Growth and Adjust Instruction in Real-Time.

- Edulastic
  - Description: Formative assessment with real-time analytics.
  - Link: https://edulastic.com/
- Schoolytics
  - Description: Dashboard that gives precise visual data on student learning.
  - Link: https://schoolytics.com/
- PowerSchool Insights
  - Description: Enterprise-level learning data insights for schools.
  - Link: https://www.powerschool.com/solutions/analytics/

Phase IV: REFLECT

Chapter 11. Continuous Quality Improvement (CQI) Made Simple with AI

Strategy 10. Refine Your Teaching With Continuous Quality Improvement (CQI), Powered by Fundamental Data.

AI Tools & Software

- BrightBytes
  - Description: Analyzes student behavior, engagement, and performance data.
  - Link: https://brightbytes.net/
- Otus
  - Description: All-in-one platform for progress monitoring and reporting.
  - Link: https://otus.com/

- Qualtrics for Education
    - Description: Surveys and learning data collection for CQI goals.
    - Link: https://www.qualtrics.com/education/

PHASE IV: REFLECT

Chapter 12. Future-Proofing Your Teaching—Staying Ahead in the Age of AI

- AI Explorations by ISTE
    - Description: Training courses to keep teachers updated on AI trends.
    - Link: https://iste.org/AI
- Future of Teaching by UNESCO
    - Description: Global insights and reports about AI in education.
    - Link: https://unesdoc.unesco.org/ark:/48223/pf0000383434
- AI4K12.org
    - Description: National guidelines for AI literacy in K–12 education.
    - Link: https://ai4k12.org/

These dashboards present all the information in one place, utilizing visual tools such as graphs, heat maps, and progress bars. This isn't about collecting more data but about making sense of it. With real-time updates, you see shifts in performance as soon as they happen, and customizable views let you track what matters most—be it reading for ELL students or math for gifted learners. Color-coded alerts draw your attention to urgent needs or achievements, so nothing goes unnoticed.

Setting up an AI dashboard is a straightforward process. Most platforms, like Classcraft Analytics or Otus, guide you through the basics. You'll typically begin by importing your class roster, often by uploading a spreadsheet or synchronizing with your school's system.

Next, select the metrics to track, such as reading fluency, math facts, participation, and homework. You can personalize alerts, for instance, receive notifications if a student's performance drops or if consistent improvement warrants recognition.

If you're tracking ELA, your dashboard may flag students who are not showing growth in comprehension. In math, you may set alerts for everyday class struggles. User-friendly interfaces and templates help you tailor the dashboard to your needs with simple drag-and-drop tools.

The key benefit comes from how you use the information. Dashboards make student growth visible in ways traditional gradebooks cannot. You'll see individual progress bars filled as students master skills, and heat maps highlight which standards are strong or need attention. Spot a student whose performance dips after a good start? That's a sign for intervention. See a student surging ahead?

Time for enrichment. At the group level, you can filter by subgroups—ELL, SPED, gifted, Title I—so patterns among those groups are instantly clear. For example, if ELL students are making progress in oral fluency but lagging in vocabulary, know where to focus support.

The visuals are actionable. If several Title I students struggle with reading comprehension, you might consider organizing a small group or connecting with families for help at home. If your gifted students plateau at "almost there," it's a cue for advanced material or enrichment. Make it a weekly habit to review your dashboard. Even better: invite students for quick **"data huddles,"** let them reflect on their progress, and set goals with your guidance. These five-minute check-ins can be valuable, like asking, ***"Jordan, your progress jumped—what changed?"*** or ***"Maria, vocabulary's tough—let's try new strategies."*** Teachers should keep a quick log of interventions and check back to see if they made a difference.

Interactive Element: Weekly Data Huddle Protocol

Step 1: Reserve ten minutes each Friday for a class or small group "data huddle."

Step 2: Highlight one key metric (reading, math, or participation) on the dashboard.

Step 3: Have students interpret their progress, share strategies, and discuss any obstacles they encounter.

Step 4: Celebrate the improvements and brainstorm the next steps for those who are still stuck.

Step 5: Teachers note planned interventions and revisit progress next week.

With these tools and routines, you'll never have to guess who needs help or wait for report cards to see results. Dashboards provide a dynamic view of your classroom, ensuring every student's progress is visible and addressed.

Setting Up AI-Driven Formative and Summative Assessments

Suppose you've ever wondered how to keep your finger on the pulse of student learning without spending every evening hunched over a pile of papers. In that case, AI-driven assessments can be a lifesaver. Let's break it down: formative assessments are those quick check-ins—quizzes, polls, exit slips—that tell you how learning is progressing day to day. Summative assessments, on the other hand, are the big finishers—unit tests, projects, and final essays that show what students have learned after a chunk of instruction. The beauty of AI in this space is its ability not only to automate but also to enrich both types. You receive instant feedback, tailored questions, and innovative scoring that adapts to your students' needs in real-time.

When you build an AI assessment, you're not just stamping out multiple-choice quizzes. These tools let you select specific academic standards or skills to target. Want to check reading comprehension for ELL students? Or maybe multiplication fluency for your math group? You can pinpoint precisely what matters. Many platforms—Edulastic is a great example—allow you to create adaptive quizzes that adjust the questions based on how each student answers. If a student breezes through the first few problems, the quiz ramps up in difficulty; if they stumble, it backs up and offers more foundational questions. This isn't just a fancy tech—it means every student gets a test that fits, whether they're working on basic skills or stretching for enrichment.

For students with accommodations or who simply think differently, AI can generate alternative assessments with just a few clicks. Perhaps you have a student who requires text-to-speech support or someone who learns best through visuals or short video responses. You can scaffold question types—such as short answer, matching, drag-and-drop, and even oral responses—so that each assessment is accessible. This flexibility is a game-changer for meeting IEP goals or language learner needs without getting bogged down in paperwork.

The real magic comes after students hit "submit." AI instantly analyzes every response and breaks down performance by skill strand or standard. You're not left sorting through a sea of numbers. Instead, you see clearly which skills were strong and which ones needed more attention. Suppose your class just finished a science unit, and the dashboard highlights strengths in vocabulary but struggles with analysis questions—you know precisely where to reteach or offer extra practice. These platforms generate student-level and class-level reports you can access at any time—no more waiting weeks for test results or flying blind on what to teach next.

AI doesn't stop at pointing out problems; it gives you concrete recommendations for what to do next. If half of your class missed the mark on fractions, the tool might suggest small-group lessons or digital practice games focused on that skill. For students who have mastered the material, AI can recommend enrichment challenges or projects to keep them engaged and motivated. Some systems even auto-generate lesson materials linked directly to assessment results, saving you planning time and helping you reach every learner.

One of the most powerful features is how easily it sets up cycles of revision and retesting, creating a true **continuous improvement loop**. Instead of a test being the end of the road, students can receive immediate feedback and have the opportunity to try again. Suppose a student falls below the mastery threshold on a **formative quiz.** In that case, the platform can automatically assign a new set of practice questions, or a review lesson tailored to their specific needs. This isn't about **"teaching to the test;"** it's about ensuring every student gets multiple opportunities to succeed and learns from their mistakes without stigma.

Encouraging students to reflect on their progress builds ownership and metacognition. After each formative assessment, many AI platforms generate self-reflection prompts for students—simple questions like "What question was hardest for you?" or "What helped you get the right answer this time?" Students jot down their thoughts in digital journals or quick response boxes. These reflections become part of their learning record and provide insight into not only what they know but also how they think about their growth.

Protocols for retesting and feedback are easy to build into your weekly routine. At the end of each week, review which students need another attempt at mastering key concepts. Use automatic re-quiz assignments for those who are below mastery—no extra setup is needed on your part. Follow up with a quick conference or a written reflection prompt so students know you're invested in their progress, not just their scores.

With these tools, assessment becomes less about catching mistakes and more about guiding each learner forward. You'll find yourself spending less time grading and more time teaching—and watching students grow not just in skills, but in confidence as well.

Translating Learning Analytics into Actionable Teaching Adjustments

You know the drill—those numbers and graphs from assessments can be impressive, even a bit daunting. But the real question is: what do you *do* with all that data? Here's where AI-generated insights change things. You're not just handed a snapshot; you're given clues for next steps. Descriptive data reveals which students struggled with fractions or which reading standard tripped up the group. But prescriptive insights go further, highlighting where to focus tomorrow's mini-lesson, or who's ready for a challenge. Rather than showing just red and green bars, the AI flags the weakest concepts and suggests which standards need revisiting.

Let's say you notice in your science analytics that students struggled with ecosystems. Instead of moving on with your planned lesson on food chains, you pause to address the gap, maybe with a hands-on activity or targeted reading. This is a targeted pivot, powered by real-time evidence, not guesswork. You might catch a pattern—maybe several students are stuck on the same math step—and run a quick intervention immediately, not days later, after grading stacks of tests.

Grouping students becomes precise, not guesswork. AI identifies those who need support and those who are ready for more. You form flexible groups—perhaps three needing reinforcement on the main idea, another cluster diving into enriched activities like advanced texts or creative projects. These aren't fixed groups.

As the data changes, so do the groupings. One week, Carlos receives help with writing; the next week, he's ahead, and someone else needs support. This fluidity lets you keep up with your students' shifting needs.

For students who are ahead, analytics open new opportunities. If someone consistently exceeds benchmarks, AI suggests **"stretch"** tasks, such as independent research, a creative writing challenge, or peer teaching. Set up learning contracts: *"Here's your data—how do you want to stretch this quarter?"* Extension menus get tailored by achievement, not by age or grade. This keeps advanced learners engaged and challenged, rather than bored.

Documenting your instructional shifts is as essential as making them. Maintain a simple digital teacher journal with notes—what you changed, and what happened as a result. Tuesday: Retaught paragraph structure. Wednesday: 85% improved on assessment." Over time, these notes become invaluable for reflecting on which strategies work. Share these reflections with your team in regular meetings.

Use easy templates: What data surfaced? What did we try? What worked? What didn't? These honest team conversations turn analytics from just numbers into collaborative action and professional growth.

Every week, the cycle restarts. Data streams in, you review it through the filter of your classroom and make quick adjustments as needed. Sometimes it means dropping a planned lesson because half the class didn't master a key concept. On other days, it's launching a spur-of-the-moment debate because comprehension was high, but critical thinking lagged. The AI is your assistant, but you're still making final decisions to fit your students' personalities and needs.

Enrichment for high performers also becomes more intentional with analytics front and center. If your strongest readers consistently excel, direct them to "stretch" prompts recommended by AI, such as analyzing author bias or designing their science experiments. For math standouts, deliver advanced problem sets or real-world challenges. Students can even propose projects tied to their strong areas, keeping them engaged and motivated.

Students should reflect, too. Regular routines where they review their progress charts and note what worked or didn't help build ownership. Over time, students come to see themselves as directing their learning path, rather than just receiving direction from teachers. These moments of reflection encourage self-awareness and resilience that last far beyond a single school year.

At team meetings, bring your outcome logs and stories of instructional shifts. One teacher's adjustment can spark an idea for a colleague facing a similar issue. Maintain easy-to-use templates so everyone can share: What was noticed? What did we do? What changed? This ongoing loop—numbers to action, action to reflection, and back—keeps teaching responsive and practical for every student, in every classroom.

Involving Parents with AI-Generated

Progress Reports and Conferences

Keeping families informed can powerfully boost student growth, but it often adds to your workload. AI-generated progress reports ease that burden by transforming complex data into clear, readable summaries of each student's strengths, needs, and next steps. Instead of overwhelming charts and numbers, both teachers and parents get a straightforward picture of student progress, sometimes for the first time. These reports don't just give grades, they highlight successes, pinpoint areas for improvement, and suggest actionable monthly goals. Best of all, they're written for families, making data accessible and meaningful.

Sending updates is flexible: choose monthly, bi-weekly, or custom frequencies to match student needs or classroom routines. AI also lets you customize the format. Some families prefer detailed breakdowns, while others want concise narrative summaries. Language barriers are less of an issue with instant translation features, making reports accessible in languages like Spanish, Mandarin, or Arabic. Now, every parent can stay informed and engaged, regardless of language or personal time constraints.

Sharing these AI-generated reports during parent-teacher conferences—virtual or in person—transforms these meetings. Rather than searching through paper files or spreadsheets, you can display dashboards that visualize mastery of standards and growth over time. Parents quickly see what's going well and what needs attention. Visuals help demystify progress, especially for parents who might otherwise feel anxious or left out.

A brief example: At a recent conference, a worried parent could instantly see her daughter's steady, real gains in math via a simple bar graph. That reassurance, impossible with old grading alone, left her not just comforted but motivated with clear ways to help at home. These tools facilitate meaningful and supportive conversations.

The real value isn't just in sharing data—it's in using it to spark collaborative solutions. Actionable insights help you and your families move from "here's what's happening" to "here's what we can try." AI tools also offer templates and scripts, making it easier to discuss results supportively: "Your child excels at main ideas but gets distracted in math. Ten minutes of reading or math games at home could help." After conferences, AI can generate summary action points for everyone—parents, students, and teachers—turning meetings into ongoing partnerships. This clarity keeps everyone accountable and ensures they work together effectively.

For smooth implementation, set routines early in the term, perhaps sending reports every other Friday, and informing families of what to expect. Ask about preferred languages at the start and keep template formats ready. Most AI platforms allow you to save and reuse what works best. And remember, celebrating even small growth with a short note ("Your hard work is showing!") can be incredibly motivating for families.

Visual Element: Sample Parent Progress Report

A typical summary might look like:

- Strength: Improved reading comprehension, now above grade level
- Growth Area: Needs work on math word problems.
- Next Steps: Try math games at home; continue nightly reading.
- Progress Graph: Shows steady improvement in reading since September.
- Translation: Provided in Spanish for family convenience

This approach truly makes parents partners in the learning process.

Bringing parents in through AI-generated reports and conferences builds trust, fosters teamwork, and supports student success. As you use these tools more, you'll notice conversations shift from "What's wrong?" to "How can we help?"—turning data into action, and action into growth.

In summary: Student data should drive relationships and authentic learning, not just fill screens or spreadsheets. **Involving families with clear, actionable information gives every** child the benefit of a supportive team. Next, we'll discuss how to use these insights to **continually improve teaching** and ensure that no student is left behind as education evolves.

"Humans and AI: together they will compose symphonies no single mind could create."

—Janelle Shane, AI researcher

Continuous Quality Improvement (CQI)

Made Simple with AI

CHAPTER ELEVEN

CHAPTER 11:

CONTINUOUS QUALITY IMPROVEMENT (CQI) MADE SIMPLE WITH AI

- ✓ **OBJECTIVE:** Educators will evaluate their instructional practice using AI reflection tools and create one improvement goal.
- ✓ **VAKT ALIGNMENT:** Visual: Visual self-assessments; Auditory: Voice notes; Kinesthetic: Peer walk-throughs; Tactual: Planning templates.
- ✓ **BLOOM'S LEVEL:** Evaluation, Synthesis
- ✓ **INCLUSIVE SUPPORTS:** Supports reflective practices for any instructional setting or learner demographic.

Implementing AI-Powered Classroom Observations for Self-Reflection

Imagine a regular classroom—not staged for evaluation but buzzing with its natural activity: a bit of noise, varied engagement, and you balancing attention among twenty-five students.

Like many teachers, you review the day in your mind:

--Did I give students enough time to respond?
--Did everyone participate?
These **reflections**, however, are often fleeting, lost by day's end.

Here's where AI becomes a helpful, impartial tool for self-reflection.

AI Tools & Software for CQI Implementation Aligned to Strategy & Phase

Phase IV: REFLECT

Strategy 10. Refine Your Teaching With Continuous Quality Improvement (CQI), Powered by Fundamental Data.

- BrightBytes
  - **Description:** Analyzes student behavior, engagement, and performance data.
  - Link: https://brightbytes.net/
- Otus
  - **Description:** All-in-one platform for progress monitoring and reporting.
  - Link: https://otus.com/
- Qualtrics for Education
  - **Description:** Surveys and learning data collection for CQI goals.
  - Link: https://www.qualtrics.com/education/

****

AI-powered observation tools shift the experience from stressful, clipboard-based evaluations to something in your control. You decide when and how to record your class—using any phone, laptop, or school device.

Modern platforms ensure privacy and confidentiality: you set up the session, control access, and secure your footage and feedback. When you record, the AI auto-generates transcripts and a digital overview of your lesson, so there's no need to comb through hours of footage unless you choose.

Here's where AI shines. It doesn't just provide transcripts; it analyzes classroom dialogue and interaction in nuanced ways.

Wait Time For instance, AI measures **"wait time"** after you ask a question—a key factor proven to boost **student engagement and thoughtful responses.** The AI calculates average pause lengths and compares them with best-practice recommendations.

You might learn that your **open-ended questions get longer wait times th**an yes/no prompts, revealing patterns you might otherwise miss.

The focus isn't on errors but on actionable insight.

Participation mapping is another AI feature. The AI tracks who speak, asking how many students contribute and how balanced the discussion is. It offers data, such as ***"Seven students spoke at least once in this session"*** *or* ***"Three students dominated small group time."***

Many platforms offer visual summaries, such as heat maps, which illustrate where you moved and which areas of the classroom received the most attention. These reveal overlooked zones or participation hotspots at a glance.

Privacy is a common concern, but this method remains teacher-controlled and confidential. You pick which lessons to record, when to delete them, and whether to share the data. Setup is a simple device placed discreetly using built-in annotation features.

AI highlights key moments, such as when discussions peak, hands go up, or an extended silence occurs. You maintain total control, ensuring feedback feels safe and is not evaluative unless you choose to make it so.

What's most powerful about this approach is its non-judgmental feedback. Instead of vague advice like **"increase engagement,"** you receive specifics: "Twelve open-ended and four closed questions were asked," or "Participation rose after **think-pair-share**."

AI can prompt reflection, such as, "Students responded more to real-world examples. How can you incorporate this more?" or "Invite quieter students to lead; only 20% spoke up today." These are suggestions, not mandates—meant to foster your growth.

Keep a weekly log after each session with prompts such as: "What surprised me about student engagement?" or "Where did higher-order thinking emerge?"

Note action steps: "Ask two new students to participate in the next class," or "Add wait time after tough questions." Over time, this creates a living growth map that documents your progress in questioning, pacing, or differentiation.

INTERACTIVE ELEMENT: SELF-REFLECTION TEMPLATE

Weekly AI Observation Reflection

1. Date / Lesson Topic:
2. What instructional strengths did I notice?
3. Where did student engagement peak? Where did it dip?
4. How was wait time distributed? Any surprises?
5. What patterns emerged in participation?
6. What is one action step I will try next week?
7. How will I measure progress (e.g., increased hands raised, more student-led questions)?

These routines are simple and require curiosity, rather than technical expertise or additional hours. Whether teaching energetic young children or skeptical teenagers, AI provides clear, unbiased snapshots for meaningful improvement.

For administrators and coaches, this method fosters trust by making professional development teacher driven. Every educator deserves feedback that's useful, actionable, and theirs to interpret and apply.

USING AI TO GATHER STUDENT FEEDBACK AND DRIVE INSTRUCTIONAL CHANGE

Students often have valuable opinions, but they aren't always easy to capture or organize into actionable insights. AI has transformed this process by quickly gathering, sorting, and highlighting feedback patterns without making students feel targeted or overwhelmed.

Instead of traditional paper surveys or reading into nonverbal cues, you can now use AI-driven surveys that students complete in minutes on any device.

Responses feed directly into a dashboard, and these surveys adapt in real time: if a student says they struggled with a project, follow-up questions pinpoint what made it challenging; if they enjoyed an activity, the next question explores what worked for them.

This branching logic yields specific, nuanced feedback, far beyond generic answers.

AI enables feedback cycles that align with your class's natural rhythms. You might create a weekly poll on favorite and least favorite activities, invite open-ended reflections like "What helped you learn this week?", or ask students for a single word to describe a lesson.

What sets AI apart isn't just collecting responses but organizing and presenting them effectively. Dashboards visualize data with word clouds, so frequently mentioned terms like "group work" or "confusing directions" stand out. Trends become obvious: an increasing interest in projects, fewer mentions of stress after you tweak homework, etc.

Getting high-quality feedback starts with well-crafted questions. Move beyond broad prompts like "Any comments?" and ask specifics: "Which activity gave you the most confidence?" or "If you could change one thing, what would it be?" Mix open responses for honesty with quick polls to keep surveys manageable.

Timing is key; frequent, short check-ins—weekly or biweekly—capture real-time impressions, unlike infrequent, end-of-term surveys. When students realize their feedback is valued regularly, they are more candid and invested.

Interpreting this wealth of data is easier with the aid of AI. The dashboard doesn't just tally numbers—it can perform sentiment analysis to identify the mood of responses and flag notable shifts. Patterns emerge: if "more group work" repeatedly appears, or if students mention "feeling lost" during a subject, you receive clear, actionable insights at a glance.

Teachers can spot early warning signs, such as mixed responses to a new project format or indirect requests for more structure from quieter students, long before these issues escalate.

However, feedback only catalyzes change if students see that their input makes a difference. Sharing "You said, I did" summaries or visible bulletin boards linking student suggestions to resulting actions demonstrates responsiveness. Whether that means more group activities, additional reflection time, or minor adjustments like more visuals for ELL students or pacing changes, showing students the impact of their feedback builds trust and engagement.

Student involvement in continuous quality improvement (CQI) can extend beyond anonymous surveys. When a trend emerges (such as requests for more partner work), invite the class to help co-create classroom norms or routines. Work together to define practical approaches for group work or feedback. AI can even assist by summarizing ideas from multiple classes or suggesting proven best practices.

Interactive Element: Sample Feedback Question Bank

- Which activity this week made you think the hardest?
- What would you change about today's lesson?
- Rate your confidence during group work (1–5).
- What would you like us to do more often?
- Was anything confusing this week? What?
- Share one thing that helped you learn today.

Rotate these questions or combine them with quick polls and open-ended prompts. For younger students or visual learners, consider pairing questions with emoji scales or requesting a short video response.

Families can also engage in this process; teachers might share results at home and invite more suggestions, which enriches the feedback pool.

These strategies don't require specialized tech skills or costly programs. Many AI tools are available for free or are integrated into existing school platforms.

The fundamental transformation is cultural: making feedback routine, acting on it promptly, and ensuring everyone sees themselves as contributors to school improvement. When students know their voices shape not only what they learn but also how they learn, their investment and engagement naturally increase.

Building a CQI Cycle—Weekly, Monthly, and Semester Checklists

CQI—Continuous Quality Improvement—sounds like something out of a corporate boardroom. Still, in reality, it's much more about finding small, honest **ways to keep getting better as an educator, parent, or learner.**

With AI's support, you can break this process down into manageable, regular habits that fit into your real life, not just your summer to-do list. You don't need a significant overhaul or a fancy system. Instead, think of it as a **steady loop: check in, reflect, tweak, and watch for growth.**

AI tools make these steps easier to **remember, track, and celebrate.**

Start with the week. Each Friday (or whenever your brain isn't fried), set aside a few minutes to review the week's **teaching and learning.** Pull up your **AI dashboard**—maybe you're using something from your school's platform or a free online tool.

Scan the Highlights:

Which lessons had the most active student participation? Did certain activities get more positive feedback? The AI will likely flag patterns you might have missed, like ***"class discussion was more even during Tuesday's debate"*** *or* ***"students requested more visuals in science."***

Use this insight to jot down one small improvement goal for next week. For example: "Incorporate visuals into math warm-ups" or "Give two extra minutes for think-pair-share." If you're pressed for time, use a weekly checklist like this:

- Run AI feedback survey.
- Check the lesson engagement heatmap.
- Set a new improvement goal for next week.

Having this routine is like setting a reminder to water your plants before they droop. Set up automatic nudges—most AI platforms can ping you with reminders or even email you a checklist, so you don't forget. If you're feeling brave, share your goal with a colleague or even with your students. That kind of accountability works wonders.

Monthly Check-Ins with AI for CQI

Monthly check-ins raise the view a bit higher. Choose one day (maybe the last Friday of the month) to review broader patterns. Review growth analytics from your AI systems to identify which students made significant jumps or who may need additional support.

Update your digital lesson bank with any freshly improved materials or add notes about what worked (or flopped).

If your platform offers it, join one AI-powered community event or forum that month—these gatherings are packed with fresh ideas and candid discussions from like-minded individuals who truly understand the topic. Here's a handy monthly checklist:

- Compile student growth analytics from the AI dashboard.
- Update the lesson bank with the best tweaks and resources.
- Attend or review one AI educator community event.

These monthly reviews are also a perfect time to document what you've tried and what's changed. Most AI platforms now offer goal-tracking dashboards where you can log progress automatically.

You might see graphs showing how many improvement cycles you've completed, or which strategies led to upticks in engagement or achievement.

These dashboards save you from scrambling at the end of the year, trying to remember what happened when.

Twice a year—or at the end of each semester—it's time to step back and synthesize everything.

Review All the AI-Generated Insights:

- --What trends stand out? Did your efforts improve reading scores for ELL students?
- --Did more group projects lead to better attendance or happier learners?

Use these patterns to set new professional goals for the next term.

Maybe you'll decide to test peer coaching strategies or push for even more student voice in lesson planning.

Here's What a Semester Checklist Might Look Like with CQI Tools:

- Review all weekly and monthly reflection notes.
- Analyze long-term growth and engagement patterns via AI analytics.
- Choose two new professional goals based on data.
- Share findings and next steps with your PLC or team.

One of the smartest moves is embedding these CQI routines into regular team meetings or PLCs (Professional Learning Communities).

Instead of everyone scrambling alone, rotate a "CQI facilitator" each month—a person who leads the group through reviewing AI summary reports, celebrating wins, and brainstorming what's next. Sharing these insights turns improvement into something social and energizing rather than just another solo chore.

AI Automated Reminder Tools

For those who want an extra nudge, use the AI tool's automated reminders. Set recurring calendar invites for CQI reviews, weekly reflections, or collaborative debriefs after big projects.

Several platforms even let you upload evidence of growth, such as a short video clip, a student work sample, or a proud parent email, so you can see your progress over time instead of just hoping things are improving.

CQI Doesn't Have to Be Complicated or Intimidating;

When scaffolded into weekly, monthly, and semester intervals, it becomes a living part of classroom culture.

Whether you're new to AI or feeling like a seasoned pro, these small habits build on each other, helping you adapt as needs change and keeping your practice fresh and responsive.

The best part is how visible growth becomes—not just for you but for every person who learns alongside you.

Networking for Growth– Connecting with Other Educators via AI Communities

One thing I've learned over the years: nobody thrives in isolation—not teachers, not parents, not even the most driven students.

We all need a place to share, laugh, vent, and learn. AI-powered educator communities are now making that connection easier and brighter than ever.

Instead of endlessly scrolling through social media or getting lost in giant email threads, you can let AI do the heavy lifting and find people for you.

Whether you're teaching second-grade phonics, high school algebra, or coordinating special programs, AI can match you with the right groups based on your interests, teaching needs, or even the types of students you support.

Imagine A Scenario Using AI

Imagine logging into a district platform or a site like Edmodo or ISTE and seeing AI personalized discussion threads handpicked by AI—no more wading through irrelevant topics or awkward introductions.

If you're curious about differentiated instruction for ELL students or want to swap strategies on project-based learning, the AI brings those conversations straight to you.

It even suggests mentors who have walked in your shoes or flags upcoming webinars on subjects that fit your current goals.

This isn't just about convenience; it's about saving your mental bandwidth for what matters.

When it comes to problems that stump you—technology glitches, tricky parent communications, lesson flops—AI-driven forums become your digital staff lounge.

You can post questions any time (even at midnight) and get responses from educators across the globe.

Chatbots embedded in these communities answer frequently asked tech or pedagogy questions instantly, so you're not waiting until the next department meeting to get help.

And there's a special magic in those spontaneous "aha!" moments when someone from another district shares a resource or a tip that solves your exact problem.

The Best Part

AI-curated book studies, collaborative webinars, and Q&A sessions keep things lively. The best part? AI sends reminders about events that mesh with your schedule and interests—no need to remember every date yourself.

If you love delving into new research or want to explore a current classroom trend, these spaces are always abuzz with fresh discussions. You can learn at your own pace—jump in live or catch up later with archived conversations and session notes.

The reach of these communities stretches far beyond local borders. Teachers now collaborate on cross-district project repositories, building massive banks of lesson plans, rubrics, and ideas for student engagement. If you have a great lesson that got kids fired up, you can upload it for AI evaluation and peer review. In return, you gain access to materials that have been tested and refined by teachers who understand the realities of absolute classroom chaos and creativity. Imagine downloading a project template that's already been adapted for SPED, ELL, or gifted students—saving hours and boosting your confidence in reaching every learner.

Of course, staying connected shouldn't mean drowning in notifications or losing precious downtime. I recommend picking just a handful of high-value groups curated by AI—your personal "go-to" list. Schedule weekly "office hours" if you want to mingle, ask questions, or catch up on threads without distraction. Set clear boundaries: mute notifications for non-urgent topics and prioritize communities where you feel genuinely supported and challenged. The point isn't to join every group but to find those spaces where growth feels natural, feedback is honest, and innovation is celebrated.

If you're new to this scene, start small. Join one or two AI-curated groups focused on issues you care about right now. Lurk a little if you want—there's no shame in reading quietly before jumping in. Share when you're ready; even a single lesson idea or question can spark valuable exchanges. As you grow more comfortable, consider submitting your lesson plans for feedback or leading a discussion thread on a topic close to your heart.

As we close this chapter, remember that growth isn't something you chase alone or achieve overnight. Next, with a supportive network powered by both human wisdom and innovative technology, continuous improvement becomes not just possible but joyful.

Futureproofing Your Teaching— Staying Ahead in the Age of AI

Chapter Twelve

CHAPTER 12:

FUTURE-PROOFING YOUR TEACHING—STAYING AHEAD IN THE AGE OF AI

- ✓ **OBJECTIVE:** Educators will explore and summarize two emerging AI trends and design a pilot activity to implement one of these trends.
- ✓ **VAKT ALIGNMENT:** Visual: Infographics; Auditory: Podcasts; Kinesthetic: EdTech exploration labs; Tactual: Printed handouts.
- ✓ **BLOOM'S LEVEL:** Knowledge, Application, Synthesis
- ✓ **INCLUSIVE SUPPORTS:** Prepares teachers and learners for innovation across all academic levels.

CURATING YOUR AI TOOLKIT—EVALUATION RUBRICS AND SELECTION GUIDES

Imagine preparing for class—dozens of tools on-screen, each claiming to save time or boost engagement. Some deliver, others disappoint, and some lose relevance as students' needs shift or technology changes. If you wonder whether you're using the best tools or overlooking something better, this chapter will help you sort meaningful solutions from fleeting trends. The aim is to create a toolkit that's practical, safe, adaptable, and truly serves your current classroom.

AI TOOLS & SOFTWARE FOR THIS CHAPTER AND PHASE

PHASE IV: REFLECT

Chapter 12. Future-Proofing Your Teaching—Staying Ahead in the Age of AI

- AI Explorations by ISTE
  - Description: Training courses to keep teachers updated on AI trends.
  - Link: https://iste.org/AI

- Future of Teaching by UNESCO
  - Description: Global insights and reports about AI in education.
  - Link: https://unesdoc.unesco.org/ark:/48223/pf0000383434
- AI4K12.org
  - **Description**: National guidelines for AI literacy in K–12 education.
  - **Link**: https://ai4k12.org/

The educational AI market is noisy—everyone touts their **"miracle"** app. To cut through the hype, you need a clear, teacher-friendly evaluation rubric. Think of it like a chef choosing ingredients: Are the tools fresh? Do they suit everyone's needs? A good rubric ensures a tool is effective, equitable, user-friendly, secure, affordable, and respectful of privacy—no jargon.

Here are five essential rubric categories:

- **Accessibility for All Learners:** Can students with disabilities, including those with IEPs or those with limited English proficiency, use it efficiently?

- **Accommodations:** Does it offer accommodations like translations, font adjustments, or text-to-speech?

- **Data Security:** Is student information encrypted, secure, and handled in compliance with privacy laws, such as FERPA? Are data practices transparent?

- **Teacher Control:** Can you customize, override, or turn off the features as needed? Does the tool allow you to review outputs before sharing them with students?

- **Cultural/Language Adaptability:** Is the content available in multiple languages? Can you adapt examples to your classroom's culture? Does it avoid bias or stereotypes?

- **Cost Transparency:** Are all costs clear upfront? Will the free version suffice, or are upgrades or hidden fees likely?

You can also find **educator-created rubrics** online that consider factors such as integration and customer support. The value in these rubrics lies not just in scoring tools, but also in prompting reflection to ensure each tool aligns with your educational goals.

Maintaining Your Toolkit

A toolkit isn't static. Treat it like digital **"spring cleaning."** At least once a year, conduct an **"AI audit."** Pull up every tool you use and ask:

- **Does it still meet this year's needs?**
- **Did the features change?**

- Is the support needed missing (translation, SEL, assessment)?
- Which tools were used regularly or caused problems?
- Did you spend unnecessary time on manual work?
- Are there supports lacking for specific student groups (SPED, ELL, gifted)?
- What did student feedback reveal—favorites or frustrations?

Keep your toolkit current and "declutter" by replacing tools that no longer fit. If you're budget-conscious, focus on multi-use applications. Don't hesitate to retire from something that's outlived its usefulness, or ***which are trying to get you to subscribe for a monthly cost.***

Context Matters

Your setting shapes your priorities. Early childhood educators may prefer voice assistants with simple commands, interactive story generators that incorporate visuals and sounds, or phonics apps.

For little ones, intuitive navigation and strong privacy controls are key. In secondary classrooms, you may lean toward AI essay graders, data dashboards, or coding bots for science classes. Specialists, such as speech therapists, will seek adaptable communication boards or robust accessibility features.

Collaborate and Validate

Don't evaluate tech alone. The best advice comes from fellow teachers with real classroom experience, rather than from vendors or marketing materials. Leverage networks: join educator review sites (like EdTech Evidence Exchange), attend virtual district "show and tell" sessions, or ask colleagues for their recommendations.

When reading reviews, be cautious of potential bias and outdated information. Focus on details about how a tool works with specific student populations—practical feedback trumps generic praise.

Interactive Element: DIY AI Toolkit Audit Checklist

Take ten minutes and use this checklist—grab your tools, device list, and maybe your tech buddy or a friend:

1. List every digital tool/app/platform used last semester.
2. Rate each 1–5 for:
   _____ Accessibility
   _____ Data Security
   _____ Teacher Control
   _____ Cultural-Language Fit
   _____ Cost Clarity.
3. Circle tools scoring three or lower in any category.
4. For circled tools: Can you address issues with settings or training, or is an
5. alternative needed?

6. Note at least one gap (e.g., better translation).
7. Set a goal: Add one new peer-reviewed tool this month.
8. Repeat the audit each term.

A curated AI toolkit isn't about collecting the trendiest tools.

In This Toolkit Book & Beyond

—It's about **supporting you and your students now**, while staying adaptable for tomorrow.

Professional Development Pathways for AI

--Workshops and Peer Learning

If you've ever tried to learn a new tech skill in the middle of a hectic school year, you know the struggle is real.

Professional development can feel like drinking from a firehose—especially when it comes to artificial intelligence.

But the truth is, you don't need to take a sabbatical or travel across the country to some teacher conference to keep your teaching sharp.

There's a wide range of flexible, bite-sized learning options available, designed to suit every comfort level and schedule. Some are as simple as a Saturday workshop, while others stretch over months and let you learn at your own pace, pajama-friendly and judgment-free.

Let's start with the quick wins. Local school districts, education cooperatives, and even your state department of education often offer one-day AI workshops as part of their PD menus.

These sessions are great for getting your feet wet, hands-on demos, Q&A with real teachers, and takeaway guides you won't just toss in the recycling bin.

Keep an eye out for virtual summits. These often include recorded sessions you can watch during a planning period or while folding laundry.

For those who crave more depth or need something on their schedule, self-paced online courses—or MOOCs (Massive Open Online Courses) are a game-changer. Platforms like Coursera and edX offer beginner-friendly courses, such as "AI for Everyone" or "AI in Education Basics."

ISTE's **"AI Explorations for Educators"** is another favorite, mixing videos, interactive tools, and real classroom examples.

Microsoft offers **free AI professional development modules tailored to teachers,** with certifications that mean something on a résumé. Many come with discussion boards, allowing you to exchange ideas with educators from every corner of the globe.

Now, if you're thinking, "I've tried online courses before and never finished," you're not alone. The secret sauce is building a personalized professional development plan that fits your real life, messy schedules, and all.

Start Small: Start SMART

Start small: Set a SMART goal (Specific, Measurable, Achievable, Relevant, Time-bound) for your AI learning.

For Example:

--"By October, I'll complete Microsoft's AI in Education module and try out one new tool with my reading group."

--Block out short windows in your calendar—maybe 30 minutes every other week or one evening a month. Track progress in a simple notebook or digital spreadsheet.

--Celebrate milestones, no matter how tiny: finishing a module, trying a new tool, or just understanding how AI spell check works.

Peer Learning

To keep things fresh and social, lean into peer learning. No one should have to figure out AI on their own.

--Start an **"AI book club"** for staff—pick an article or video each month and meet over cookies or coffee to discuss what clicked (or didn't).

--Pair up with a **"tech buddy"**—maybe someone who loves gadgets with someone who'd rather write lesson plans by hand—and commit to learning together.

--Rotate who leads quick **"AI Lunch & Learn"** sessions, where teachers showcase their favorite tools in short bursts—no pressure to be perfect, just share what works (and what doesn't).

Learning Circles

Learning circles are especially powerful when you invite everyone: classroom aides, counselors, even students who might surprise you with their tech savvy. As your confidence grows, you may want to pursue further advanced badging or certification.

There's power in modeling curiosity and grit for learners of every age. When you make growth visible—on a classroom wall or in a staff newsletter—you permit others to explore right alongside you.

Cost-Effective AI for Low-Budget Schools—Grants, Open Source, and DIY Solutions

Even though I have tried to offer you as many free AI tools as possible in this toolkit — with proper precautions outlined in my disclaimer —there are other opportunities if you wish to delve deeper for personal growth and learning, to reach every student to a **greater metacognitive height.** (including those identified as **gifted and talented,** as well as other **advanced learners**). When money is tight and your classroom wish list reads like a fantasy novel, it's easy to think AI is out of reach.

However, I've seen firsthand that even the most budget-conscious educators can introduce innovative technology to their students, with a bit of creativity, a dash of persistence, and a sprinkle of the attitude that says, **"let's make it work."**

This project will allow every student, regardless of background or ability, to access personalized support through **free AI translation apps**. Our goal is to bridge learning gaps and give every child a fair shot."

Not every solution requires outside funding. There's a whole universe of free and open-source AI tools built with teachers in mind.

These platforms level the playing field, allowing every classroom to benefit from powerful analytics, adaptive lessons, and creative content generation—no credit card required.

OpenSALT provides open-source learning analytics dashboards, enabling you to **track student growth** without incurring expensive licensing fees. For **literacy and math, Khan Academy Kids** stands out as a treasure trove of adaptive activities, stories, and games—completely free, with no hidden paywalls.

Google Read Along turns any mobile device into a reading buddy that listens and gives gentle feedback as kids practice aloud. If you're supporting **multilingual families** or students with limited internet access, **Google Translates** offline features and **Microsoft's Immersive Reader** can help bridge communication gaps without incurring any costs.

Resourcefulness often beats big budgets.

Teachers in various settings have found innovative ways to stretch their resources.

In some classrooms,

--**Smartphones** double as document scanners—snap a picture of student work, upload it to an AI grading app, and get instant feedback even if you don't have laptops for everyone.

--**Downloadable lesson templates** allow you to plan offline and upload when your bandwidth allows.

--If your school has only one computer lab or spotty Wi-Fi, **rotate small groups through digital stations** while others work on hands-on projects or paper-based activities.

After-School Computer or AI Club

Consider forming an after-school AI Club that could be enjoyable for both students and teachers. Students have access to complimentary online resources and platforms.

AI Translation Features

Another teacher in rural Maryland helps her multilingual families utilize free AI translation features to stay informed.

She sent home videos with auto-generated subtitles in two languages and used WhatsApp voice notes that were translated into the parents' native tongues. It wasn't flashy, but it worked, and students saw their home cultures valued right alongside their lessons.

Every tool mentioned here has one thing in common: it removes barriers instead of creating them.

Whether you're applying for grants, hunting down free apps, or reimagining what you already have on hand, the real magic comes from adapting ideas to fit your unique context.

Th most impactful tech

The most impactful tech in any school is the one used thoughtfully by teachers who know their students' stories and are willing to try something new.

The path may zigzag and require some creative thinking. Still, with resourcefulness and community support, AI can become an everyday part of learning for everyone, regardless of zip code or funding stream.

Vision for the Future—Empowering Students as AI-Ready Lifelong Learners

If you've ever wondered about the future your students will inherit, you're not alone. The future isn't some distant concern—it's already at our classroom doors.

Artificial intelligence is changing more than just how we grade or automate; it's shaping the very world our students will live in, work in, and vote in.

Your role matters more than ever: you're not just teaching content but equipping every student—no matter their background or ability—with the skills and mindset to thrive, question, and help shape an AI-powered future.

AI literacy is no longer a luxury.

AI literacy is no longer just for "computer nerds" or a luxury—it's a new baseline for equity and opportunity. If students don't understand how AI works, they risk being excluded from vital conversations about jobs, privacy, fairness, and democracy.

Agency is at stake: when only a few students understand AI, power stays with a select group. But when every child learns to talk to, question, and even build AI tools, they gain a say in shaping their world. Actual civic participation will increasingly depend on recognizing when AI is assisting us and when human oversight is needed. Your classroom is where that balance begins.

Being "AI-ready" means building key competencies. Students should understand what AI is (and is not), how it learns from data, and where it appears in daily life—from YouTube recommendations to virtual assistants. Prompt engineering is a foundational skill; students need to ask clear, precise questions and refine prompts to get valuable results from AI tools. Ethical use is crucial: spotting bias, questioning AI outputs, and understanding privacy concerns before sharing data. Data interpretation matters, reading charts, spotting skewed numbers, and asking what's missing are vital skills. Creative problem-solving with technology—like brainstorming ideas with AI or building simple tools—puts students in an active, empowered role.

Teachers have successfully introduced "**Introduction to AI**" mini-curricula in grades 4–8. Start simple: **What is AI? How do computers detect patterns?** Use **hands-on** activities where students "**train**" each other with if–rules, then compare with how computers learn. Practice prompting AI chatbots: let students try different question phrasings, analyze the responses for accuracy, and discuss any unusual results that arise. For older students, stage debates on topics like: ***"Should schools use AI to flag mental health issues in student writing?"*** or ***"Is it ethical for news outlets to use AI-generated articles without clear labeling?"*** These conversations foster critical thinking and help students see themselves as active participants in shaping technology, rather than just passive users.

Embedding these skills in everyday learning—not just special units or after-school clubs—makes AI literacy truly stick. Cross-curricular projects boost engagement and relevance. In ELA or journalism, students can use AI-driven fact-checkers to spot fake news. In science, challenge groups to design an AI tool that addresses a community problem, such as tracking water usage or sorting recyclables. Let kids pitch ideas, sketch prototypes (even just on paper), and reflect on what worked or surprised them. Encourage student-driven inquiry: What bugs them about existing tech? Can they design a more innovative chatbot for homework or a playground rule translator for new classmates?

Of course, **this only works if teachers model curiosity and adaptability.** Be open with students: **"I'm still learning this too—let's figure it out together."** Build routines for ongoing learning—maybe explore a new AI app, article, or topic every Friday as a class. Use simple, regular reflection—like a journal prompt: "**What's one thing I learned about AI this month?**"—to help everyone notice growth over time. Celebrate small successes, such as debugging a program or asking a truly creative AI question, to build confidence for students and teachers alike.

You don't need to have all the answers to help students become future-ready. What matters most is welcoming questions and giving space for experimentation and learning from mistakes. By cultivating these habits in your classroom, you spark a ripple effect that lasts far beyond grades or the latest technology. In a world where AI influences everything from shopping to elections, confident and critical thinkers who act ethically will be more important than ever.

As this chapter closes, remember that preparing students for an AI-driven world isn't just about coding or using new gadgets; it's about building timeless human skills. Stay curious, stay open, and keep inviting students to join you as you navigate whatever comes next.

Conclusion

Well, here we are—the end of our journey through "THE AI TOOLKIT FOR TEACHERS MADE EASY." If you've made it this far, first of all, give yourself a high-five or a pat on the back. (No, really—go ahead, I'll wait, my dear fellow educator.) I am proud of you for trying and not giving up on this AI technology "thingy."

Continue to explore various tools and software available now (and new ones that will become available in this ever-changing world) to build your "very own toolkit" for your classroom of unique students.

Accolades:

You stuck with me through the twists and turns of AI movement, strategies, lesson planning, grading, critical thinking, VAKT learning styles, SEL, phases of instruction and teaching, and more.

I trust that you are feeling more confident and perhaps even eager to explore additional tools in your toolkit.

The mission of this book, from page one to the very last template in the appendix, has been simple: to give you, the tools and encouragement to use AI in a way that saves time, fits real classrooms, and makes sure every student—gifted, struggling, learning English, or working with an IEP—gets what they need to thrive.

I've always believed teaching is about opening doors, not building walls, and this toolkit is meant to be your key ring.

It doesn't matter if you still call your laptop "the computer thingy" or if you're the go-to tech buddy (or tech wizard) in your hallway or your school. There's a spot in my heart for you.

"AI offers more opportunities to identify minor issues before they escalate into larger problems."

Let's take a quick victory lap and revisit the 10 AI strategies you have mastered and now have in your toolkit:

1. Simplify lesson planning—Let AI do the heavy lifting on frameworks so that you can spend more time with your students than with your keyboard.
2. Personalize learning paths—No more "one size fits all." Use AI to help each student take steps at their own pace, in their way.
3. Automate grading and feedback—Say goodbye to endless piles of papers and hello to evenings spent doing things you enjoy.
4. Differentiate instruction—Tweak lessons quickly for SPED, Title I, ELL, ESOL, and gifted students, without rewriting everything from scratch.
5. Engage students creatively—gamify, create simulations, and let AI spark and cultivate curiosity in your classroom.
6. Enhance critical thinking—Use AI to build Socratic seminars, debates, and rich reflection moments, so your students become thinkers, not just memorizers.
7. Integrate SEL and wellness—Make space for emotions and well-being with AI-powered check-ins and empathy-building activities.
8. Collaborate globally—Break down the walls of your classroom and connect students across towns, countries, and continents, all with the help of AI tools and translation.
9. Assess growth using AI analytics—See every student's progress, strengths, and needs at a glance, and adjust your teaching on the fly.
10. Refine instruction through CQI—Reflect, adjust, and celebrate your growth, using AI to make minor, steady improvements all year long.

ALL OF THESE STRATEGIES FIT NEATLY INTO THE FOUR PHASES OF REAL TEACHING LIFE:

I. PLANNING,

II. TEACHING,

III. ASSESSING, And

IV. REFLECTING.

You don't have to do them all at once. Pick the one that makes the most significant difference for you and your students right now, and trust that the rest will fall into place as you explore more tools in depth and build confidence.

I want to emphasize this: This toolkit is for everyone. I wrote it with you in mind, whether you teach kindergarteners who can't tie their shoes or high schoolers who can code circles around the rest of us. Every template, guide, and lesson were designed to be adaptable—to reach students who have learning differences, who speak different languages at home, who might feel invisible in traditional classrooms, or who need just one spark to ignite their curiosity and confidence.

THE BIGGEST TAKEAWAY I HOPE YOU CARRY WITH YOU IS THIS:

AI is here to help, not to replace you. It's the teaching assistant who never gets tired, but **you are still the heart and soul of your classroom.** You bring empathy, judgment, creativity, and relationships that no machine can match. Practical, ethical, and student-centered AI is not only possible, but also doable for every teacher, regardless of their tech skills or school funding. Tiny steps are fundamental steps. Small wins add up. Nobody expects you to turn into **"Super Teacher 2.0"** overnight.

Here's something I learned after decades in education: Curiosity and courage go further than perfection. Experiment. Use the sample lesson plans, VAKT activities, IEP templates, and study guides as your launching pad. Share your discoveries and tech failures with colleagues. Celebrate your successes, and don't be afraid to tweak or toss what doesn't work for your students.

As you use these tools, remember the ethical side of tech. **Guard your students' privacy** like you would your house keys. Watch for bias and speak up if you spot something off. Maintain open communication with parents and families. Let them see the good, the honest mistakes, and the ways you adapt. That's how trust grows.

SELF-PRAISE

It is essential to recognize and appreciate your achievements and contributions. Every time you try something new, whether it's an AI-powered quiz, a global classroom exchange, or a simple chatbot for homework help, you're modeling the future for your students.

You're showing them it's okay to be a beginner, to get creative, to ask questions, and never to stop learning. That's how we raise kids who are ready for whatever the world throws at them—AI and all.

If you ever feel stuck or just want to swap stories, don't go it alone. There are educator communities online (including ours at www.driven2succeed.com), and there's a whole world of teachers cheering you on. Use AI Brain Teasers and Glossary to keep your skills sharp. Reach out, connect, and keep the spirit of innovation alive.

So, here's my **last challenge for you: pick one AI strategy from this book—just one. Try it out in your classroom. Reflect on what changes.**

Notice the time you get back, the spark in a student's eyes, or the way a lesson **"clicks."** Then, share that win. You never know who you might inspire.

Above all, believe in yourself. You've got what it takes to lead, to adapt, and to help every student **believe, achieve, and succeed**—no matter what the future brings.

I believe in you—Blessings on your next step in your teaching adventure.

BONUS SECTIONS

- Study Guide & Workbook (For Each Chapter)
  • Acrostics, Crossword Puzzles, Graphic Organizers, Brain Teasers
- Glossary of Terms
- Recommended Free AI Teaching Tools & Software Links
- Brain Teasers & Study Guide
- Lesson Plan Templates
- IEP Samples
- Master Index
- Answer Keys per chapter!
- Resources & References (APA Style)
- Index

GLOSSARY

These definitions and examples are used to reflect the teaching pedagogy that AI is being utilized to personalize learning, save time, and empower every student to succeed:

Definitions and Examples

- Accommodations: Adjustments made to teaching methods or testing requirements to help students access the curriculum more effectively.
- ADHD (Attention Deficit Hyperactivity Disorder): A neurodevelopmental disorder that affects attention, impulsivity, and activity levels.
    - Example: Students may benefit from structured learning environments.
- AI (Artificial Intelligence): Computer technology programs that can "think" or solve problems in ways that mimic human intelligence, simulating human intelligence processes, such as learning, reasoning, and self-correction.
    - Example: AI tutors provide personalized assistance to students.
- Automated Grading: Software that checks student work and assigns scores.
- Bloom's Taxonomy: A framework for categorizing educational goals by cognitive levels from simple recall to higher-order thinking.
    - Example: Assess the level of awareness your students are presently performing (Knowledge, Comprehension, Analysis, Application, Synthesis, and Evaluation)
- Critical Thinking: The ability to think clearly and rationally about what to do or believe.
    - Example: Analyzing information during discussions.to using AI to master a subject using metacognition
- Differential Learning: Adjusting or adapting instruction and various teaching methods to accommodate different learning styles and needs.
    - -Example: Write a personalized individual education plan (IEP) for all students, not just for special education students, using AI.
- Digital Assistant: A computer program designed to help you manage tasks.
- ELL/ESOL (English Language Learner/English for Speakers of Other Languages): Students whose first language is not English.
    - Example: Tailored instruction helps these students acquire English proficiency.
- Graphic Organizers: Visual tools to help students organize information, such as charts or mind maps.
    - Example: Teach students to create their graphic organizers, such as Venn diagrams, to demonstrate their understanding of a particular concept (e.g., compare and contrast).

- **IDEA: (Individuals with Disabilities Education Act)** was first signed into law in 1975, as the Education for All Handicapped Children Act (EHA), mandating that all students with disabilities (SWDs) receive free, appropriate public education(FAPE) in the least restrictive environment (LRE) possible.
- **IEP (Individualized Education Program/Plan)**: A document developed for each public-school child who needs special education, outlining specific learning goals and accommodations. A student's FAPE and LRE are established through a process that results in an IEP for each student.
- **Metacognition:** Awareness and understanding of one's thought processes.
  - **Example:** Students reflecting on how they solve problems.
- **Natural Language Processing: How computers understand and generate human language (like when ChatGPT writes text).**
- **SEL (Social-Emotional Learning):** Programs that focus on developing students' emotional intelligence and social skills.
  - **Example: Partner and** Group activities that foster friendship and teamwork.
- **Special Education Student:** Students who require specialized educational services due to their disability/disabilities.
  - **Example:** Students with learning disabilities may need differentiated instruction.
- **Title 1 Program**: A federal program providing financial assistance to schools serving low-income students, aiming to ensure all children meet state standards.
  - **Example:** Federal Title 1 funds are used to supplement the regular school's budget to provide extra resources for the low-income/low-performing school students
- **VAKT Learning Styles (Visual, Auditory, Kinesthetic, and Tactile) Learning Styles** that illustrate different preferences in learning.
- **ISTE: International Society for Technology in Education (ISTE)** is a nonprofit organization that sets the standard for how technology should be used effectively in teaching and learning.

✅ **WHY IT MATTERS:** It's most famous for the ISTE Standards, which are widely used in schools, districts, and universities to guide curriculum and professional development, help educators integrate technology responsibly and creatively, and build students' digital skills and citizenship.

How it relates to my AI TOOLKIT FOR TEACHERS book: When I reference ISTE, I am aligning my work to recognized, research-based best practices for digital learning.

Official ISTE Website to Learn More: ISTE.org

NOTE: Many AI grants, professional development programs, and accreditation processes aim to align with ISTE Standards.

SUMMARY:

AI TOOLS, SOFTWARE, WEBSITES, LINKS, & RESOURCES:

AI Free Tools organized by The Four Instructional Phases (PLAN; TEACH; ASSESS; & REFLECT), The 12 Chapters, and the 10 Strategies *(Explore these at your leisure to keep abreast of AI and other Technology to Help Us work Smarter & more efficiently—Not Harder)* –

(Disclaimer Note: Skip Ads that pop up in most free links, apps, and software. Although many free links offer advanced plans for a monthly fee, there are sufficient free links available for your students. Try other links under the relevant topic. If any of the links are not active, try cutting and pasting the link into your browser.)

PHASE I: PLAN

Chapter 1. Demystifying AI for Every Educator

AI Tools, Software, & Resources:

- AI for Education by ISTE *(see the details about ISTE in the Glossary):*
  - Description: Intro course helping educators understand AI in simple terms.
  - Link: https://iste.org/AI
- Google's AI for Education
  - Description: Free AI tools and courses from Google tailored for teachers.
  - Link: https://edu.google.com/
- Microsoft Learn AI for Educators
  - Description: Microsoft's free educator training for AI integration in classrooms.
  - Link: https://learn.microsoft.com/en-us/training/educator-center/

Phase I: PLAN

Chapter 2. Building a Safe & Ethical AI Foundation

Strategy 1: Simplify lesson planning with AI tools.

AI Tools & Software:

- Common Sense Education – AI Guidelines
  - Description: Guides educators and schools on the ethical use of AI with vetted resources.
  - Link: https://www.commonsense.org/education
- AI Ethics Curriculum by MIT Media Lab
  - **Description:** Curriculum for teaching students about AI and responsible use.
  - **Link:** https://aieducation.mit.edu/

Phase I: PLAN

Chapter 3. AI for Effortless Lesson Planning

Strategy 2: Personalize Learning Paths for Every Student.

AI Tools: & Software

- LessonPlans.ai
  - **Description:** Auto-generates standards-aligned lesson plans from simple prompts.
  - **Link:** https://www.lessonplans.ai/
- TeachMateAI
  - **Description:** Tool with 60+ generators to simplify teacher tasks.
  - **Link:** https://www.teachmateai.com/
- Eduaide.ai
  - **Description:** Personalizes lessons and quizzes with AI-driven suggestions.
  - **Link:** https://eduaide.ai/
- Curipod
  - **Description:** Interactive slides from one-sentence prompts. Great for students engagement and planning personalized learning.
  - **Link:** https://curipod.cocham/
- Diffit
  - **Description:** Adapts any article or topic to various reading levels instantly.
  - **Link:** https://www.diffit.me/

PHASE II: TEACH

Chapter 4. Differentiating with AI—Personalized Learning for Every Student

STRATEGY 3: AUTOMATE GRADING & FEEDBACK TO SAVE YOUR SANITY.

- Khanmigo
  - **Description:** An AI tutor that helps students while teachers monitor progress.
  - **Link:** https://www.khanacademy.org/khan-labs
- Socratic
  - **Description:** Students take a photo of their homework to get explanations.
  - **Link:** https://socratic.org/
- ScribeSense
  - **Description:** An AI tool that helps grade handwritten or typed assignments.
  - **Link:** https://www.scribesense.com/
- ELSA Speak
  - **Description:** Helps ELL students improve speaking and pronunciation skills.
  - **Link:** https://www.elsaspeak.com/en/

PHASE II: TEACH | PHASE III: ASSESS

Chapter 5. Streamlining Grading & Feedback With AI

STRATEGY 4: DIFFERENTIATE INSTRUCTION, EVEN WHEN YOUR CLASS FEELS LIKE A JIGSAW PUZZLE.

- Gradescope
  - **Description:** Grades exams and homework using AI to save teacher time.
  - **Link:** https://www.gradescope.com/
- Writable
  - **Description:** Provides structured feedback on student writing drafts.
  - **Link:** https://www.writable.com/

- Formative
    - **Description**: Creates quizzes and real-time insights from student answers.
    - **Link**: https://goformative.com/
- **ZipGrade**
    - **Description:** Scans and grades multiple-choice tests using a smartphone.
    - **Link**: https://www.zipgrade.com/
- **Germini**
    - **Description**: Germini is currently under construction but aims to create active learning experiences that encourage student engagement and participation.
    - **Use**: Intended for educators to facilitate dynamic interactions in the classroom, fostering a love of learning in every student.
    - **Link: htps://germini.org**
- **Kahoot**
    - **Description:** Kahoot! is a game-based learning platform that enables educators to create engaging quizzes and learning games to assess student understanding.
    - **Use:** Teachers utilize Kahoot! to engage students in learning through interactive quizzes and games that foster participation and collaboration.
    - **Link** https://kahoot.com
- **Nearpod**
    - **Description:** Nearpod is an interactive classroom tool designed to engage students through live polling, quizzes, and multimedia presentations.
    - **Use:** It helps teachers deliver high-quality instruction by promoting active learning and differentiation, allowing them to instantly see students' understanding through feedback.
    - **Link**: https://nearpod.com
- **Pear Deck**
    - **Description:** Pear Deck is an interactive presentation tool that empowers teachers to create engaging lessons with real-time feedback and student interaction.
    - **Use:** It is widely used to facilitate active learning and assess student understanding through collaborative activities and instant polling.
    - **Link**; https://peardeck.com

- Socrative
    - **Description:** Socrative is an assessment tool that allows teachers to create quizzes and engage students with activities that provide real-time feedback.
    - **Use:** It's utilized to monitor student progress and understanding instantly, making classroom management and feedback simpler and more efficient.
    - **Link:** https://socrative.com
- Edulastic
    - **Description:** Edulastic is an online assessment platform that enables teachers to create, administer, and analyze assessments while providing instant feedback to students.
    - **Use:** It is designed to align with standards and helps track student progress over time, making it easy to personalize learning tasks.
    - **Link:** https://edulastic.com
- **Formative**
    - **Description:** Formative is an assessment tool that enables teachers to create interactive assignments and quizzes for real-time tracking of student understanding.
    - **Use:** Teachers can use Formative to collect instant responses, assess student work in real-time, and adapt instruction on the fly.
    - **Link:** https://goformative.com
- **Flipgrid**
    - **Description:** Flipgrid is a video discussion platform that enables teachers and students to engage in discussions through video responses.
    - **Use:** It's widely used to promote student voice and collaboration, allowing learners to share ideas and perspectives through short videos.
    - **Link;** https://info.flipgrid.com

PHASE II: TEACH

Chapter 6. Creative Engagement—Supercharging Student Motivation with AI

STRATEGY 5: ENGAGE STUDENTS CREATIVELY—BRING SOME SPARK BACK TO LEARNING.

AI Tools & Software:

- Book Creator with AI
  - Description: AI-enhanced tool for students to create digital books.
  - Link: https://bookcreator.com/
- Storybird
  - **Description:** Let students write and illustrate AI-enhanced stories.
  - **Link:** https://storybird.com/
- Animoto
  - **Description:** Turn lessons into engaging videos with drag-and-drop AI.
  - **Link:** https://animoto.com/
- Tome AI
  - **Description:** Storytelling platform enhanced with AI design elements.
  - **Link:** https://tome.app/

PHASE II: TEACH

Chapter 7. Fostering Critical Thinking and Higher-Order Skills with AI

STRATEGY 6: STRENGTHEN CRITICAL THINKING, NOT JUST ROTE ANSWERS.

- Kialo Edu
  - **Description:** Structured platform for guided debates and argumentation.
  - **Link:** https://www.kialo-edu.com/
- Perplexity AI
  - **Description:** AI research assistant for deeper inquiry-based learning.
  - **Link:** https://www.perplexity.ai/
- ThoughtExchange
  - **Description:** Facilitates group thinking and higher-order conversations.
  - **Link:** https://www.thoughtexchange.com/

PHASE II: TEACH

Chapter 8. Integrating SEL & Wellness through AI

STRATEGY 7: INTEGRATE SOCIAL-EMOTIONAL LEARNING AND MENTAL WELLNESS.

AI Tools & Software:

- Peekapak
  - **Description:** SEL curriculum infused with AI personalization options.
  - **Link:** https://www.peekapak.com/
- Along
  - **Description:** AI check-ins and reflections to promote student well-being.
  - **Link:** https://www.along.org/
- MindSpark Learning
  - **Description:** Professional development for AI and SEL fusion.
  - **Link:** https://mindspark.org/

PHASE II: TEACH

Chapter 9. Global Collaboration—Connecting Classrooms with AI

STRATEGY 8: COLLABORATE WITH CLASSROOMS (AND EXPERTS) AROUND THE GLOBE.

AI Tools & Software:

- Empatico
  - **Description:** Connects classrooms globally for empathy and cultural learning.
  - **Link:** https://empatico.org/
- PenPal Schools
  - **Description:** Global collaboration platform for project-based learning.
  - **Link:** https://www.penpalschools.com/

PHASE III: ASSESS

Chapter 10. Data-Driven Assessment & Growth Tracking with AI

STRATEGY 9: UTILIZE AI ANALYTICS TO EVALUATE GROWTH AND ADJUST INSTRUCTION IN REAL-TIME.

- Edulastic
  - **Description:** Formative assessment with real-time analytics.
  - **Link:** https://edulastic.com/
- Schoolytics
  - **Description:** Dashboard that gives precise visual data on student learning.
  - **Link:** https://schoolytics.com/
- PowerSchool Insights
  - **Description**: Enterprise-level learning data insights for schools.
  - **Link:** https://www.powerschool.com/solutions/analytics/

PHASE IV: REFLECT

Chapter 11. Continuous Quality Improvement (CQI) Made Simple with AI

STRATEGY 10: REFINE YOUR TEACHING WITH CONTINUOUS QUALITY IMPROVEMENT (CQI)–POWERED BY FUNDAMENTAL DATA.

AI Tools & Software

- BrightBytes
  - **Description:** Analyzes student behavior, engagement, and performance data.
  - **Link:** https://brightbytes.net/
- Otus
  - **Description:** All-in-one platform for progress monitoring and reporting.
  - **Link:** https://otus.com/
- Qualtrics for Education
  - **Description:** Surveys and learning data collection for CQI goals.
  - **Link:** https://www.qualtrics.com/education/

PHASE IV: REFLECT

Chapter 12. Future-Proofing Your Teaching—Staying Ahead in the Age of AI

AI Tools & Software

- AI Explorations by ISTE
  - **Description:** Training courses to keep teachers updated on AI trends.
  - **Link:** https://iste.org/AI
- **Future of Teaching by UNESCO**
  - **Description:** Global insights and reports about AI in education.
  - **Link:** https://unesdoc.unesco.org/ark:/48223/pf0000383434
- AI4K12.org
  - **Description:** National guidelines for AI literacy in K–12 education.
  - **Link:** https://ai4k12.org/

****

This Toolkit Book IS NOT a textbook; however, as forward-thinking teachers, not just for today, but are preparing for the fast-changing tech world, I still wanted to share some interesting professional articles, resources, AI Educational Articles, Microsoft AI Education, Common Sense Media AI Ratings, & EdTech Reviews of AI Software.

Stay Updated & Watch Reviews by Microsoft AI Education, Common Sense Media AI Ratings, & EdTech Reviews of AI Software: https://edu.google.com/

Before I list the other AI Tools, Software, and Resources, I wanted to give you a preview of one of the **latest AI Tools: Kuripath,** which is not readily found. However, it is similar to Curipod and MagicSchool, two of the teachers' favorites for personalizing learning. The reviews on this relatively new tool have gone well (see info below)

🔍 WHAT IS AI KURIPATH?

(HTTPS://WWW.KURIPATH.COM).

Kuripath (also stylized as KuriPath) is an AI-driven platform designed to **enhance personalized learning and help teachers save time while improving student achievement.**

While it's newer and not as widely known as tools like Curipod or MagicSchool, it's gaining traction, especially among teachers seeking simplified planning and differentiation.

KURIPATH – FEATURES & DESCRIPTION

FEATURE	DESCRIPTION
AI Lesson Generation	Allows teachers to generate lesson plans aligned with standards instantly.
Customization	Lessons and activities can be modified to suit different grade levels or IEP needs.
Student-Centered	Provides AI-suggested strategies for meeting individual learning styles.
Data-Driven	Provides analysis tools to track student progress and adjust instruction accordingly.
Differentiation	Tailor materials based on reading level, language needs, and learning goals.
Time-Saving Templates	Includes pre-built templates for IEPs (Individualized Education Plans), RTI (Response To Intervention), 504 Plans (Need Related Services/Accommodations), etc.

****

✓ How Teachers Use Kuripath (https://www.kuripath.com).

- Plan Smarter, Not Harder: Teachers input standards or objectives, and Kuripath generates lessons, activities, and assessments.
- Support Inclusion: Helpful for SPED, 504, ELL/ESOL, and Gifted learners via differentiated options.
- Document easily: Generate accommodation plans, lesson reflections, and parent communication logs with a few clicks.
- Integrate with LMS: Often works with Google Classroom and other Learning Management Systems.

"Humans and AI: together they will compose symphonies no single mind could create."

— Janelle Shane, AI researcher

Other AI Tools, Software, and Resources

Again, explore these at your leisure to stay up to date on AI and other technologies that help us work smarter and more efficiently, not harder.(Cut & Paste URL if it does not populate)

- Edutopia: "Artificial Intelligence in Education." - https://www.edutopia.org/
- Educause AI: "7 Things You Should Know About Artificial Intelligence in Higher Education." https://www.educause.edu/
- Getting Started with AI: A Beginner's Guide for Educators: https://www.aiforeducation.io/blog/getting-started-with-ai
- Overcoming the Fear of Technology: https://theowlteacher.com/fear-of-technology/
- Ten Myths About Artificial Intelligence in Education https://scholarworks.waldenu.edu/cgi/viewcontent.cgi?article=1508&context=hlrc
- AI Readiness Framework: https://www.aiedu.org/s/ai_readiness_framework_district_leaders.pdf
- AI and the Law: What Educators Need to Know – Edutopia: https://www.edutopia.org/article/laws-ai-education/#:~:text=FERPA%20(Family%20Educational%20Rights%20and,data%20with%20third%2Dparty%20providers.
- AI Policy Template for Schools (2025 Update): https://ictevangelist.com/ai-policy-template-2025/
- 5 Strategies for Integrating AI Ethics in the Classroom: https://www.numberanalytics.com/blog/classroom-ai-ethics-strategies
- Thinking About Equity and Bias in AI - Edutopia https://www.edutopia.org/article/equity-bias-ai-what-educators-should-know/#:~:text=A%20proactive%20approach%20is%20required,between%20students%20of%20diverse%20backgrounds.
- AI for Educators | MagicSchool; https://www.magicschool.ai/
- AI Lesson Planner for American Teachers: https://www.kuraplan.com/country/united-states
- Eduaide.ai Transforms Lesson Planning With Best AI Tools: https://www.revoyant.com/blog/eduaide-ai-transforms-lesson-planning
- 10 AI Tools Teachers Should Explore This Summer: https://shakeuplearning.com/blog/10-ai-tools-teachers-should-explore-this-summer/
- Use AI for differentiated instruction – Microsoft 365: https://www.microsoft.com/en-us/microsoft-365-life-hacks/everyday-ai/use-ai-for-differentiated-instruction
- The Top 10 Adaptive Learning Platforms https://www.thirdrocktechkno.com/blog/top-10-adaptive-learning-platforms-in-2024/

- Supporting ELL Instruction with AI: Ideas for Educators: https://www.colorincolorado.org/article/supporting-ell-instruction-ai-ideas-educators#:~:text=Platforms%20like%20adaptive%20learning%20software, scaffolded%20activities%2C%20and%20draft%20feedback.
- Report - Off Task: EdTech Threats to Student Privacy and Equity https://cdt.org/insights/report-off-task-edtech-threats-to-student-privacy-and-equity-in-the-age-of-ai/
- Top 24 AI Grading Tools for Teachers for Efficient: https://www.essaygrader.ai/blog/ai-grading-tools-for-teachers
- AI for Impact: https://www.writable.com/ai/
- Transform grading into learning with Gradescope: https://www.turnitin.com/products/gradescope/
- Using AI to Reduce Bias in Education: Promoting Equity with ... https://www.linkedin.com/pulse/using-ai-reduce-bias-education-promoting-equity-tools-ruci-m-a-m-ed y9mpe#:~:text=AI%2Dbased%20grading%20platforms%20use,playing%20field%20for%20all%20students.
- Four Tools to Help Gamify Your Classroom https://www.middleweb.com/23970/four-tools-to-help-gamify-your-classroom/
- Labster | Virtual Labs for High Schools and Universities: https://www.labster.com/
- The Role of AI in Revolutionizing Personalized Learning: https://cloudgate.ie/blog/the-role-of-ai-in-revolutionizing-personalized-learning#:~:text=AI%2DPowered%20Virtual%20Tutors%20and%20Chatbots&text=They%20can%20explain%20concepts%2C%20answer,Duolingo%20(language%20learning)
- Using AI Tools in the Classroom https://edu.google.com/intl/ALL_us/ai-resources-in-education/
- Parlay Ideas | AI-Powered Class Discussions: https://parlayideas.com/
- How Students Can Use AI in Project-Based Learning https://www.edutopia.org/article/how-students-use-ai-pbl-units/
- Building Metacognitive Skills Using AI Tools to Help Higher ... https://dialnet.unirioja.es/descarga/articulo/10068132.pdf
- Free Tools and AI Resources to Support Social-Emotional ... https://blog.tcea.org/free-tools-and-ai-resources-to-support-social-emotional-learning/#:~:text=1.,generator%20creates%20detailed%20integration%20plans.
- Mood Meter: https://hopelab.org/stories/mood-meter/
- Leveraging AI to improve student mental health | DATIA K12 https://www.datiak12.io/leadership/oped/article/15743522/leveraging-ai-to-improve-student-mental-health#:~:text=AI%2Dpowered%20mental%20health%20monitoring,saving%20lives%20through%20early%20intervention.
- Predict student learning styles and suitable assessment ... https://www.sciencedirect.com/science/article/pii/S111086652400032X?dgcid=rss_sd_all

- I Translator: Breaking Language Barriers - Murf AI https://murf.ai/blog/ai-translator#:~:text=A%20popular%20and%20user%2Dfriendly,and%20compatible%20with%20various%20devices.
- Getting Started with Global Connections in the Classroom https://www.ksuiteach.org/blog/global-collaboration
- Top 6 AI Project Management Tools for ... https://instituteprojectmanagement.com/blog/the-6-best-ai-project-management-tools-to-help-you-succeed/
- Tackling digital safety challenges to create a safer world https://www.weforum.org/stories/2025/01/tackling-emerging-harms-create-safer-digital-world-2025/
- 5 AI Student Progress Tracking Tools Improving ... https://www.kangaroos.ai/blog/ai-student-progress-tracking-tools-improving-student-development/
- Assessment Settings Overview: https://dsst.zendesk.com/hc/en-us/articles/4412306773403-Assessment-Settings-Overview
- Using-Artificial-Intelligence-to-Transform-Curriculum-for-Gifted-Students-and-Professional-Development-for-Teachers - https://nagc.org/blogpost/2061726/509113/
- Parent Teacher Conference Template: https://www.notably.ai/ai-templates/parent-teacher-conference-summary
- AI Coach for Teacher PD | Instructional Coaching https://www.edthena.com/ai-coach-for-teachers/
- AI feedback tool improves teaching practices | Stanford Report https://news.stanford.edu/stories/2023/05/ai-feedback-tool-improves-teaching-practices
- Twelve Tips to Guide the Use of Data with Generative AI for ... https://www.aamc.org/media/82576/download?attachment
- ISTE | Artificial Intelligence in Education: https://iste.org/ai
- Rubric for Evaluating AI Tools for Schools https://www.controlaltachieve.com/2024/04/rubric-for-evaluating-ai-tools-for.html
- 40 AI Tools for Teachers, Educators, and Classroom (Free) https://ditchthattextbook.com/ai-tools/
- ISTE | Artificial Intelligence in Education: https://iste.org/ai
- Classroom Grants for Teachers: Secure STEM Resources https://nextwavestem.com/stem-grants-for-teachers

---

"To understand the future of AI, look not at what machines can do, but at what they enable humans to become."

— Kai-Fu Lee, AI Superpowers

REFERENCES

Here are some recent and recommended AI/EdTech References: Relevant resources, sources, books, articles, and other links on AI that should provide a more solid foundation of information for teachers.

Books & Articles On AI

Anderson, M. (2023).
Learning to Learn with AI:
Teaching Strategies for the Digital Classroom.

Darling-Hammond, L. (2023).
Equity & Technology in Education:
A Framework for the Future.

McKay, D. & Wallace, C. (2024).
The AI Educator's Field Guide: Prompting, Planning, and Practicing.

International Society for Technology in Education (ISTE). (2024).
AI in Education Toolkit

Anderson, J., & Rainie, L. (2022).
AI and the Future of Education.
Pew Research Center.

Vuorikari, R., & Castaño Muñoz, J. (2023).
Digital Education at School in Europe.
Publications Office of the European Union.

Woolf, B. P. (2023).
Building Intelligent Interactive Tutors:
Student-Centered Strategies for Revolutionizing e-Learning. Morgan Kaufmann.

Articles on AI:

Critical Thinking Assessment in Higher Education
https://www.tandfonline.com/doi/full/10.1080/10447318.2025.2499164

UNESCO: Artificial Intelligence in Education:
Challenges and Opportunities for Sustainable Development."
https://unesco.org/

OECD: AI and the Changing Education Landscape.
https://www.oecd.org/education/

BONUS TREAT FOR YOU & YOUR STUDENTS: MY INTERACTIVE WORKBOOK (ABBREVIATED)

My Interactive Workbook from My *THE TOOLKIT FOR TEACHERS MADE EASY* book is designed for your students to plan, learn, practice, reflect, and set goals (under your wise guidance) – all with a bit of AI magic sprinkled in.

To Our Students: Each activity was designed to engage your eyes, ears, hands, and mind (VAKT), and to challenge different thinking skills from Bloom's Taxonomy (remembering facts, understanding concepts, applying skills, analyzing and evaluating information, and creating new ideas).

By completing these activities, you aren't just learning about your class subjects – you're also seeing how AI can be a helpful partner in your education, and how you can think in new ways utilizing AI and other technologies—and more importantly, how to THINK and APPLY new information with (or without) technology.

LET'S DON'T RELIE TOO HEAVILY ON AI OR OTHER TECHNOLOGY THAT WE FORGET TO USE OUR OWN MIND/BRAIN—THE BEST COMPUTER THERE IS!

You + AI Creativity + Curiosity = SUCCESS! (All Working Together)

This Workbook Features:

- ✓ The 10 AI-powered Actionable Measurable Teaching Strategies
- ✓ Activities that are grouped into the **Four** instructional Phases of Learning (& Teaching): **PLAN II. TEACH; III. ASSESS; and IV. REFLECT**).
- ✓ Each Strategy includes differentiated tasks for **Grades 4–6, Grades 6–8,**
- ✓ And **Grades 9–10**, Plus a **bonus all-grades** activity (from remembering to creating),
- ✓ Engage different learning styles (**Visual, Auditory, Kinesthetic, Tactile** – VAKT).
- ✓ Interactive, printable workbook pages that reflect my vision.
- ✓ Skill-based content using Bloom's Taxonomy and VAKT learning styles.
- ✓ Each phase will include three exercises—one for grades 4–6, one for grades 6–8, and one for grades 9–10, plus one bonus mixed-grade activity.
- ✓ All exercises are designed for workbook interaction: puzzles, diagrams, matchups, and editable, printable sections.

Students: Get Ready to Solve Puzzles, Discuss, and Create as You Explore How AI Can Enhance Learning!

Happy Learning & Succeeding!

✓ **Completing** this Workbook is Proof of How Much You've Grown. You Have Master Several of These Free AI Tools.

CONGRATULATIONS!

THE AI WORKBOOK
STRATEGY TIME

10 SIMPLE STRATEGIES

TO SAVE TIME, PERSONALIZE LEARNING,

Empower Every Student to SUCCEED

-Even If You're Not Tech-Savvy

PHASE 1: PLAN

STRATEGY ONE

LESSON PLANNING: LIGHTNING LESSON PLANNER (PHASE: PLAN)

AI helps teachers (and students) plan quickly and efficiently.

ACTIVITY:

****

GRADES 4–6 (DEMONSTRATING SKILL USING: KINESTHETIC & VISUAL):

Imagine you're planning a **birthday party** with help from a friendly AI.

Below are steps the AI suggested, but they're out of order!

Number the steps 1-3 in the correct planning order (1 = first, 3 = last):

- ___ **Buy a cake** for the party.
- ___ **Send invitations** to your friends.
- ___ **Play games and music** at the party.

(Hint: What should you do first?)

After ordering, **draw a small 🎂 next to Step 1 (tactile drawing action)** – the first thing to plan. This activity utilizes sequencing *(Bloom's Comprehension/Understanding*) and encourages you to move ideas around (kinesthetic learning).

****

GRADES 6–8 (APPLYING | VISUAL & LOGICAL):

Say, you're planning a science project on growing plants, with AI suggesting tasks.

The AI gave these steps jumbled up:

a. Experiment and observe the results.
b. Formulate a hypothesis (guess what will happen).
c. Gather materials (seeds, soil, water).
d. Conclude from your results.

Re-order these steps in a logical scientific process (number 1–4).

Write the correct sequence in the blanks:
This requires critical thinking (Bloom's *apply*) to organize a plan.
Check your answers –

Did you put "gather materials" before experimenting? (That should be Step 2!) Excellent job planning visually on paper.

1: (b);
2: (c);
3: (a);
4: (d).

****

GRADES 9–10 (ANALYZING | WRITING & VISUAL):

Plan a **community service event** (e.g., a neighborhood clean-up) with an AI's guidance.

The AI asks you three planning questions.

Write a brief answer for each:

1. **"What is your goal for the event?"**
   – My goal is to ______________________________.
2. **"What tasks or supplies will you need?"**
   – I need ___________________________________.
3. **"When will it happen?"**
   – It will be on ______________________________.

By answering, you're fleshing out a plan (Bloom's ***analysis of*** the situation and needs).
This written exercise is **tactile and visual.**

For an extra **auditory element, read your plan answers aloud** to yourself or a peer – does it sound ready?

****

🎯ALL GRADES BONUS – AI PLAN & IMAGINE:

AI can help plan any project, big or small.

Think of a **dream--a project** (a school play, a game tournament, etc.) that you'd love to do.

On a separate sheet, sketch a quick poster or mind map of your project idea **(visual or kinesthetic learning style).**

- Include at least ***three things*** an AI could help plan (e.g., scheduling, ideas for activities, a checklist).
- Make it colorful! (Yes, Color IT!)
- Afterwards, ***share*** one idea from your plan with the class **(auditory)** – you're practicing communicating a plan. **Great planning, future project leaders!** 🎉

STRATEGY TWO

JUST-RIGHT RESOURCES (PHASE I: PLAN)

AI helps tailor learning materials to different levels – everyone gets what they need!

ACTIVITY:

****

GRADES 4–6 (REMEMBERING | VISUAL & TACTILE):

AI can **simplify hard words** to make them easier to understand. davidsondavie.edu

Match each **difficult word** on the left with its **easy meaning** on the right by drawing a line (or writing the letter):

- **enormous** — A. very tired
- **weary** ______ B. very big
- **leap** ________C. jump

Which matches did you make?

- ***Enormous*** **= B (massive),**
- ***Weary*** **= A (very tired),**
- ***leap*** **= C (jump).**

If you got them, give yourself a ★!

This matching **exercise (Bloom's 'remember' and 'understand' new words) demonstrates how AI can transform complex vocabulary into simpler terms**.

It's a **visual** exercise – you see connections, and **tactile** as you draw lines.

(If you're unsure, **ask your "Tech Buddy"(classmate) or say the words aloud** together – **auditory** learning helps too!)

****

GRADES 6–8 (UNDERSTANDING | READING & AUDITORY):

Below are two explanations of the same concept.

Circle the one that is easier to understand (the one you'd give, maybe, on a 5th-grade level):

- **Original:** ***"Photosynthesis is the process by which chlorophyll-containing organisms convert light energy into chemical energy to fuel their activities."***
- **Simplified:** *"**Photosynthesis** means **plants use sunlight to make food** for themselves."*

You likely circled the second sentence – it's much simpler!
AI can take complex scientific texts and simplify them like this.

(For an **auditory** twist, **read both explanations out loud** – notice how the simplified one is clearer when heard.) By evaluating clarity, you exercised Bloom's **Knowledge** (***understanding***).

Think about it: which words did the AI simplify? " *Convert"* became "**use**," and "chemical energy to fuel their activities" became "**make food**." That's *just-right* wording! 👇

****

Grades 9–10 (Applying & Creating | Writing & Visual):

Practice being an AI simplifier: **Rewrite the following formal sentence in a friendly way**.

Original: *"The **precipitation** was **sporadic**, so the team **postponed** the outdoor experiment."*

- Your simpler version: "It was ____ on and off, so the team ____ the outdoor experiment."

Fill in the blanks with easier words (hint: *precipitation* means rain; *sporadic* means off-and-on; *postponed* means delayed). One example answer: "It was **raining** on and off, so the team **delayed** the outdoor experiment." Great!

Your version uses everyday words.

This writing task (**Bloom's *apply/create***) and reading comprehension check your understanding.

Share your sentence with the class **(auditory)** – did others choose similar, simple words? AI will be proud of your paraphrasing skills. ✨

****

🎯 All Grades Bonus – AI Thesaurus Game:

Let's play with **synonyms**!

Below are words to Unscramble.

Unscramble the easy synonym for each:

- **difficult** → *unscramble:* (hint: not easy)
- **hard** → RAHD
- **tiny** → *unscramble:* SMALL – **small**.
- **Intelligent** → *unscramble:* SMART – **smart**.

How many did you get right?

This fun puzzle taps into memory **(Bloom's '*remember*')** and **tactile** skills.

(writing the answers).

It's like you became a mini-AI thesaurus, finding the right word! If you enjoy this, try making your own synonym scrambled.

AI makes ***differentiation*** **(adjusting to level)** a game – and you just did it, too! 🎲✅

STRATEGY THREE

CREATIVE AI CO-CREATOR (PHASE II: TEACH)

AI can generate stories, drawings, and more inspiring creativity in the classroom.

ACTIVITY:

****

GRADES 4–6 (CREATING | VISUAL & TACTILE):

AI can assist in creating images and characters. Imagine an AI drew this friendly **robot** for you! **Color the robot** (use your crayons – visual and tactile fun) and give it a name. As you color, think about what helpful thing this robot might do in class (maybe it passes out papers or tells jokes?). Share your idea with a friend (auditory). By coloring and imagining, you're using creativity (Bloom's *create*) just like an AI artist would. Great job bringing the AI's drawing to life!

****

GRADES 6–8 (CREATING / WRITING & AUDITORY):

Time for an AI-powered **Mad Libs** story! **AI Madlib Generator:** Create hilarious and unique stories using the MadLib AI-powered generator. (www.aiadlibs.com)

Get creative by simply:

--filling in words based on various categories,
--entering a prompt, and
--watching your story come alive.

Practice Exercise: Use the AI Madlib Generator:
Tool: www.aiadlibs.com:

Fill in the blanks with your ideas, then read the funny story you "co-wrote" with AI:

- An AI wants to write a fairy tale. It asks you for a **hero's name** (***),***
- ***a place (***), and
- a **magical object** (______).
- Now read your story: *"Once upon a time, **a hero went to a place** with a special **object**.*
- **When students read their creative stories aloud (auditory) to the class or their tech buddy, watch the laughter explode.**
- *Together, they used it to save the day. Everyone cheered for **the hero, the great hero of the place**!"* 🪄

How does your story sound?

-- Swap stories with a classmate and
--Read each other's story aloud (or silently) for laughs (auditory).

This activity (Bloom's ***Create***) demonstrates how AI can help initiate a story, and *you* add creative details to it. The writing **(Tactile)** and listening (Auditory) combine to create a full VAKT experience. Enjoy the funny, mystery, and silly stories you and AI created together.

GRADES 9–10 (EVALUATING | READING & WRITING):

AI wrote a short poem, but it made a mistake.
Read the poem and correct the error:

"Roses are red,
Violets are blue,
Sugar is salty,
And so are you."

Did you spot the odd word?
The AI mistakenly wrote, *"Sugar is salty"*
Cross it out and write **"sweet"** above it.

Now the poem makes sense: sugar is sweet! By editing the AI's work, you practiced critical evaluation (Bloom's E*valuation*) and writing. **Discuss**: Why might AI make a silly mistake like that? (Maybe it mixed up facts, spellings—Naes, etc.

NOTE: You need to review, proof, and edit whatever you research (prompt) AI to do. AI makes mistakes also!) This shows the importance of *human* review. You just teamed up with AI to improve a creation – good job, editor!

****

🎯 All Grades Bonus – AI Creativity Word Search:

AI can create **art, stories, music, and other forms of content.**

Below is a word search grid with some creative things AI can make.

- **Find and circle the words ART, CODES, POEM,** and **SONG** hidden forward or downward.
- This puzzle builds your visual scanning skill (and patience!).
- Have fun – you're discovering what AI might create: Mathematica;
  Copy Edit
  C O D X S
  X X X O X
  P X E M X
  X X X G X
  A X T X X
- *Words to find:* **CODE, SONG, POEM, ART. (Did you find them all? ☺)**
- Word searches engage **visual skill** and **problem-solving (Bloom's A*pplication*).**
- **Creativity** can be hiding in unexpected places, just like these words.

Bonus challenge: Can you find the word **"CODE"** in the grid? (It's in the top and third rows!) Imagine these words are ideas with which **AI** can help.

The possibilities are endless – and you've had a taste of them. 🎨♪ You keep creating stories, editing them, and reading them aloud—even if it's only to yourself. Who knows, one day you could be writing movies, screenplays, and speeches for officials or leaders (Heck, you could very well be one of those leaders yourself!)

STRATEGY FOUR

AI STUDY BUDDY (PHASE II: TEACH)

AI can act like a personal tutor or learning buddy, answering questions and giving hints.

ACTIVITY:

****

GRADES 4–6 (REMEMBERING | READING & DECISION-MAKING):

Some questions are better for a **teacher (T)** and some for an **AI (A)**.

Write T or A next to whom you'd ask:

1. "Can I go to the bathroom?" – **(T)**
2. "What's 5 + 7?" – **(A)** (An AI can solve math facts!)
3. "How do you spell **elephant**?" – **(A)**
4. "May I sit next to my friend?" – **(T)**

How did you decide? Questions about personal or classroom rules (1 and 4) are for the human teacher. Simple factual questions (2 and 3) – an AI can answer those. Great job sorting!

This activity (Bloom's *remember/understand* who to ask) was visual (you read and wrote answers). For a touch of auditory learning, **discuss your answers with the class** – did everyone get that an AI can't give bathroom passes?

😄 Now you know when an AI study buddy can help and when to turn to a teacher.

GRADES 6–8 (ANALYZING | CRITICAL THINKING):

AI can answer questions, but is it always **correct**? Below are questions with an AI's answer. **Circle** whether each AI answer is True (correct) or False (incorrect):

- Q: "Who wrote *Hamlet*?" AI: "**William Shakespeare.**" → True / False? **True.** ✅
- Q: "The Pacific is the largest ocean. True?" AI: "**No, the Atlantic is bigger.**" → True / False? **False.** ⊘ (The Pacific is the largest davidsondavie.edu.)
- Q: "What is 7 × 8?" AI: "**54**." → True /False? **False.** ⊘ (7×8 = 56.)

How did you do? If you marked the Shakespeare answer True and caught the two AI errors, great! 🎉 **Lesson:** Always **double-check AI's answers**.
This exercise (Bloom's *analysis*) required you to critically evaluate the information. It's visual (reading the Q&A) and a bit tactile (circling answers).

If you weren't sure, it's a reminder that even an AI study buddy can make mistakes, and why your knowledge (or asking a teacher) is essential.

Grades 9–10 (Applying | Problem-Solving & Auditory):

Getting good help from an AI is about **asking the right questions**. Which of these questions would likely get a better, more specific answer from an AI?

Put a checkmark (✓) next to the better question in each pair:

- ____ A) "Help me with science." **OR** ____ B) "What causes **thunder** and **lightning**?"
- ____ A) "I need math help." **OR** ____ B) "How do I solve an equation like 2x + 5 = 15?"

You should have checked **B** for both sets. 🎯 Specific questions (B) yield more precise answers, while vague ones (A) might confuse the AI.

This was a thinking exercise (Bloom's *apply* knowledge of clarity) – you used your understanding to predict outcomes. For kinesthetic learning, **stand up and give a thumbs-up** for each B option you chose (a little movement makes learning fun!).

Discuss: Why are B questions better? (They give details the AI can focus on.) Now you're learning the skill of **prompting** – asking effective questions, which is key when using AI as a study buddy. 💡

All Grades Bonus – Partner Quiz (Auditory & Social): Pair up with a classmate and take turns being the "AI study buddy." **Quiz each other** on any topic you both know (e.g., times tables, state capitals, vocabulary).

For example, ask, "What's the capital of France?" and let your partner answer. Give hints if needed, just like an AI would.

Then switch roles. By quizzing and explaining, you're reinforcing knowledge (Bloom's *'remember' and 'apply') and practicing effective* communication (auditory and social learning).

High-five your partner – you both just became AI-style study buddies for each other!

STRATEGY FIVE

GLOBAL AI CONNECTOR (PHASE II: TEACH)

AI tools can connect classrooms worldwide, translating languages and fostering cultural exchange.

ACTIVITY:

GRADES 4–6 (UNDERSTANDING | VISUAL & AUDITORY):

People say "**hello**" around the world in many languages. AI translators can help us understand each one.

Exercise: MATCHES: Draw a line to match each greeting to its language or country:

• Hello –	A. Spanish (Spain, Mexico)
• Hola	B. Japanese (Japan)
• Konnichiwa	C. French (France)
• Bonjour	D. English (USA, UK…)

✅Answers: Matches:

• Hello	– D (English),
• Hola	– A (Spanish),
• Konnichiwa	– B (Japanese),
• Bonjour	– C (French).

✅ How many did you get?

Now **say each word aloud** (auditory) and see if a friend can guess the language – it's fun!

This matching game (Bloom's ***understanding** of* cultural facts) shows how AI can help bridge languages.

The **visual** map in your mind grows when you learn new greetings.

AI apps can instantly translate "Hola" to "Hello."

Cool, right? 🌍🤗 Now *try other words and translate them into other languages,*

****

GRADES 6–8 (APPLYING | READING & WRITING):

Let's see AI translation in action.
Below are three Spanish words and their English meanings, but they're mixed up!

Match each Spanish word to its English meaning by writing the correct letter:

- gato – (a) house
- libro – (b) cat
- casa – (c) book

****

Answers: ***gato*** **= cat (b),** ***libro*** **= book (c),** ***casa*** **= house (a).** Great! You used logic (and perhaps recognition of *"casa," which means* "house") to decode the language.

This exercise (Bloom's: applying knowledge across languages) is visual and somewhat analytical. If you weren't sure, imagine asking an AI translator – it would give those answers in a snap.

For an extra challenge, use each Spanish word in a simple English sentence (e.g., "The *gato* is sleeping" – you mixed languages, but an AI could still figure it out!). You're on your way to being a global communicator. 🌐✍

****

GRADES 9–10 (ANALYZING | WRITING & REFLECTIVE):

Imagine your class is doing a video exchange with students in another country. AI will translate between languages in real-time.

Write one question you would like to ask the overseas students about their life or school. Then write **one sentence about your life** that you'd like to share with them. For example:

- Question for them: "What is a popular sport at your school?"
- Something about me: "We have football games every Friday night at our school."

This task (Bloom's *analysis* and *creation*) prompts you to think about cultural exchange. It's primarily written (tactile) but also reflective – you considered what you want to learn (inquiry) and what to share. AI would translate your question and answer so both classes understand – how amazing is that? After writing, **discuss your question/answer with a peer** (auditory) and see what they wrote. You've just prepared to connect globally, with a little help from AI. 🌏

****

🎯 ALL GRADES BONUS – Global Greetings Poster (Visual & Cultural):

****

The world is connected by communication. In the image above, each speech bubble could represent a "hello" from a different country! **Color each bubble** a unique color (visual creativity).

Inside each bubble, you can even **write "hello"** in a language you know or find interesting (Hola, Ciao, Ni hao, etc.). As you color, imagine kids in those countries saying hello back to you. AI translation tools can turn those hellos into your language and vice versa, helping everyone understand. When finished, display your colorful greeting poster. It's a small world after all – and now you have a rainbow of hellos to prove it!

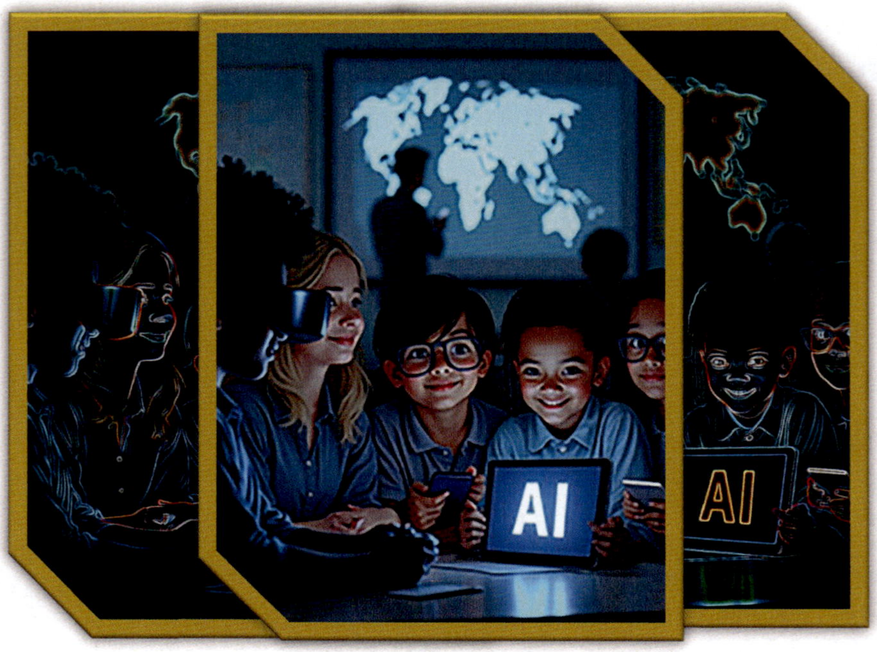

STRATEGY SIX

AUTOMATIC: GRADER & FEEDBACK (PHASE III: ASSESS)

AI can grade quizzes and provide instant feedback, helping you learn from mistakes more quickly.

ACTIVITY:

****

GRADES 4–6 (APPLYING | WRITING & VISUAL):

Here's a sentence a student wrote:

"My **favorite** subject is math." An AI spell-checker flagged a mistake.

Can you find it? **Underline** the misspelled word and **write the correct spelling** above it.

The word "favorite" is missing a letter – it should be "**favorite**."

✅ Great catch! By correcting it, you acted like an AI-powered spell checker (Bloom's *applies* knowledge of spelling).

This was tactile (writing the correction) and visual (seeing the word change).

Pat yourself on the back – you helped improve the "graded" work! Wouldn't it be neat if an AI could put a spellcheck squiggly line under "favorite" for you? Now you know what to fix.

****

GRADES 6–8 (ANALYZING | CRITICAL THINKING):

Let's simulate AI grading.

Mark each answer as correct (✔) or incorrect (✗) using the answer key:

- Q1. *6 + 4 = 10* – Student wrote **10**. **Key: 10.** → ✔ Correct.
- Q2. *Spell "friend"* – Student wrote "**friend.**" **Key: "friend."** → ✗ Incorrect.

How did you mark them?

--Q1 matched the key, so it's correct.
--Q2 had the letters mixed up, so it's incorrect – an AI (or you as grader) would mark it wrong and perhaps suggest "Check spelling of friend."

This exercise (**Bloom's *Analysis*** – you compared answers to a key) is what AI auto-graders do at lightning speed.

It was also **visual and tactile** as you read and marked.

Fun fact: Some AI graders even give partial credit or specific feedback, not just ✓/✗.

For Q2, what feedback would you give?

Maybe "Swap the middle letters." Great job thinking like a grading AI! ✓✗

Grades 9–10 (Evaluating | Writing & Reflective):

Read the short claim and the AI's feedback, then **improve the answer**:

- **Student's answer:** "Exercise is essential."
- **AI feedback:** *"Why is it important? Please **provide one reason** or example."*

Now it's your turn to respond to the feedback.

Complete the sentence with a reason:

"Exercise is important because ______________________."

For example, you might write "because **it improves your health**." Excellent – you've added supporting detail!

- This is precisely how AI feedback helps you *strengthen* your answers (Bloom's E*valuation* and improvement).
- The task was writing **(tactile)** and **thinking critically** about *why*.
- Compare with a classmate: Did you give a similar reason **(health, strength, mood)**?
- There's no single right answer, but the key is that you **responded to feedback**.
- That skill will make both you and any AI assistant happy – it means you're learning and refining your work!

****

🎯 ALL GRADES BONUS – "Great Job" Sticker (Creative & Visual):

****

Even AI graders can be encouraging – think of the fun messages you get like **"Good work!"** or digital stickers.

Design a sticker or stamp.

--That you'd love to see on your graded work.

--In the space below (or on another sheet),

Draw a fun "AI feedback" sticker –

--Maybe a computer high-fiving you,

--A robot giving a thumbs-up, or

--A badge that says "AI-approved ✅."

--Make it as creative and colorful as you like.

--This artsy task taps your creativity (Bloom's *Create*) and visual skills.

--When you're finished, share your sticker design with classmates – it's like giving each other positive AI-style feedback!

--And remember, whether it's from your teacher or an AI, **feedback is friendly** – it's there to help you grow.

STRATEGY SEVEN

ADAPTIVE PRACTICE & QUIZZES (PHASE III: ASSESS)

AI can adjust practice questions to your level, giving easier or more complicated questions as needed.

ACTIVITY:

****

GRADES 4–6 (UNDERSTANDING | CHOICE & KINESTHETIC):

Choose your challenge!

AI tutors often let you select a difficulty level that suits you, then adjust if you need more challenge. Try it now:

- Level 1 (basic): 3 + 4 =?
- Level 2 (advanced): 13 + 29 =?

Solve the level that feels right for you.
If Level 1 was easy-peasy,

Go ahead and solve Level 2 as a bonus.
Write your answer:
Level 1 = 7;
Level 2 = 42. Did you get them? (Level 2 answer: 42.)

By choosing your level:

- You practice self-assessment (Bloom's) to understand your comfort level.
- This also got you writing numbers (Tactile) and thinking actively.
- If you tried both levels, awesome!
- If you started easy and then leveled up, that's adaptive learning in action.

****

GRADES 6–8 (APPLYING | PROBLEM-SOLVING):

--Sometimes AI provides hints when you're stuck. Try this two-step math problem. Cover the hint at first, and use it only if needed:

- Problem: *Calculate (28 ÷ 4) + 3 =?*
- *(Hint: 28 ÷ 4 = 7)*

Solve it yourself first. 28 ÷ 4 = 7, then 7 + 3 = 10.
Did you get 10?
If so, you didn't need the hint–high–five! ✋

- If you used the hint, that's okay – that's what it's there for.
- You applied math skills (Bloom's **Application**) and learned how a helpful prompt can guide you.

This was **visual and logical. --How did using a hint feel?**

AI might do that: "Here's a tip, now try again." It's like a safety net.

- Now, create **your hint** for a question:
- Think of a question you know how to solve and write a hint for it (this makes you understand it).

For Example, Q:
9 × 5, Hint: "Remember 9 × 5 is half of 9 × 10."
You're learning to think like a helpful AI tutor!

****

GRADES 9–10 (EVALUATING | SELF-ASSESSMENT):

An AI tutor might ask you how confident you feel to tailor the practice.

Check one:

- I want **Easy** questions first.
- I'm ready for a **Medium** level.
- Give me the **Hardest** problems!

Now explain **why** you chose that level:

____

For example, "I chose Hard **because** I've mastered this topic" or "Easy, because I'm just starting to learn this.")

- This meta-cognitive exercise (Bloom's E***valuation of*** **your** ability) makes you **reflect** on your learning process **(WHAT you learned, HOW you learned it, and HOW you can teach your knowledge or skill to someone else).**
- That reteaching process (metacognition) is demonstrating that you know that you know a subject.
- When it's mostly writing **(tactile)** and thinking, there's no wrong answer – just being honest is key.
- In class, discuss your choice **(auditory)** if you'd like: it's interesting to see different comfort levels.

🌟 The Cool Part? **AI** would take your input and **adapt it,** accordingly, providing easier questions to build your confidence or tougher ones to challenge you. You just practiced self-advocacy in learning, which is exactly what adaptive tech encourages!

****

🎯 ALL GRADES BONUS – Pattern Challenge (Visual & Logical):

****

--AI loves patterns and might adjust difficulty with puzzles.

--Try these two pattern challenges (choose one or both!):

- **Bnm,kjkilo89;pChallenge A (easy):** 2, 4, 6, __ ? (__).
- **Challenge B (hard):** 1, 1, 2, 3, 5, __ ? (__).

🎯**Fill in the blanks:**

Challenge A is counting by 2s, so the following number is **8** davidsondavie.edu.

Challenge B is the Fibonacci sequence (each number is the sum of the two before it), so the following number is **eight** as well!

Did you get one or both?

If you get an A, try B next time as a challenge.

If you got B – wow, you tackled a hard one! 🎉

This puzzle sharpened your analytical skills **(Bloom's *analysis*)** and was visual (seeing the sequence).

It's a taste of how an AI might throw in a brain teaser adapted to your level.

Stretch your brain like this often, and you'll be pattern-proficient! ⊠✨

STRATEGY EIGHT

DATA-DRIVEN PROGRESS TRACKER (PHASE III: ASSESS)

AI can track your learning data (scores, progress) and help you see how you're improving.

ACTIVITY:

****

GRADES 4–6 (UNDERSTANDING | VISUAL):

Last week, you got **2 out of 5** questions right on a quiz. This week, you got **4 out of 5** right.

Did you improve? Circle:

YES / NO. Obviously, **yes!** 😊 You've doubled your score.

Color in 4 stars below to show this week's result (and two stars for last week, for comparison):

Previous week: ⭐ ⭐
This week: ⭐ ⭐ ⭐ ⭐

Which has more stars? This week! You just read a simple data report (**Bloom's *understanding* progress**). It was both **visual (represented by stars as a bar graph)** and tactile (involving circling and coloring).

***How* did you improve?**

- **Maybe more studying, or an AI practice game, helped?**
- **Discuss with your teacher one thing that helped you go from 2 to 4.**
- **Keep it up – those stars show you're shining brighter each time!**

GRADES 6–8 (ANALYZING | MATH & PREDICTION):

AI can help calculate and predict progress.

- **Look at your quiz scores over 3 weeks: 60%, 70%, 80%.**
- **They're going up!**

📈 Answer these questions:

- What was your **highest score? 80%**.
- If the trend continues, what might your score be in Week 4? **90%** (perhaps!).

Great job analyzing the data **(Bloom's *analysis* trend)**.
It's like you drew a mental line graph of improvement.

Now think:

- *Why* might your scores be rising?
- Write one reason:

"I started **using an AI flashcard app** for practice," or
"I **spent more time** studying each week."
By **reflecting** on **cause and effect**, you're learning from data.

This exercise involved both **numerical (tactile** calculation) and **visual (imagining** the chart) components.
Real **AI dashboards** do this – they'd show you climbing from **60% to 80%** and predict **90% (even 95 %**) next.

And you just did it yourself! Keep aiming high – you're on the right track—Who knows, you could reach 100%--just keep Believing, Achieving, and SUCCEEDING. YOU CAN DO IT! I Have FAITH In YOU!

****

GRADES 9–10 (EVALUATING | CRITICAL THINKING & WRITING):

An AI tutor notices a pattern:
You do better on assignments submitted at 7 P.M. than ones submitted at midnight.
It shows your average grades: **85% (evening)** vs **70% (late night)**.

Give one possible explanation for this pattern:

____

(e.g., "I'm less tired in the evening, so I focus better," or "I rush when I do work late at night.")

- There's no single correct answer – this is about **evaluating** your habits **(Bloom's *evaluate*** your learning data).
- By writing a reason, you practiced honest **reflection** and **critical thinking**. In the future, **AI** might *recommend* **"Try to do homework earlier."**

Based on your explanation, do you agree with that advice?
Circle one: YES / NO.

- **If Yes**, you've learned from the data and decided to change a habit – that's growth! 📈
- **If not,** there may be another factor the **AI** didn't know about (and you can note what it is).
- **Either way,** you engaged with data like a scientist in your learning. 🔍✨ **GREAT JOB!**

All Grades Bonus – Mood Tracker (Reflective & Visual):
Not all progress is measured by testing scores.

- Learning is also about mindset.
- AI mentors sometimes gauge your **mood or confidence** over time.
- Let's do a simple mood data log for this week.

For each day of the week,

- Draw a **smiley face ☺ (great)**,
- **Neutral face 😐 (okay)**, or
- **Frown ☹ (tough day)**

****

To **reflect on** how you felt about your learning that day:

- Monday: ___________
- Tuesday: ___________
- Wednesday: _________
- Thursday: ___________
- Friday: _____________

Take a look at your week of faces – this is your **emotional** learning data!
Do you see more smiles as you get closer to Friday? Is there a pattern (perhaps mid- week, you feel tired)?

Share one observation with a friend or the teacher.
This activity combines art **(drawing faces and tactile elements)**, self-awareness, and a bit of data **analysis (Bloom's '*analyze*** your feelings' trend).

- It shows that **AI tracking isn't just about grades**
- It can help you (and your teacher) notice how you *feel* during learning.

And a happy learner often learns more!

So, keep those spirits up, keep smiling (intentionally), and use the data to take care of yourself, **leading to better class performance and ultimately, better grades...** ☺💖

STRATEGY NINE

REFLECTIVE LEARNING WITH AI (PHASE IV: REFLECT)

AI can prompt you to reflect on what you learned, solidifying knowledge and encouraging growth.

ACTIVITY:

****

GRADES 4–6 (REMEMBERING | WRITING & VISUAL):

REFLECTION TIME!

Tool: davidsondavie.edu
Think back on today's lesson.

- *What is one new thing you learned?*
- **Finish the sentence** (in writing or drawing):

"Today I learned about ________________________________."

For example, you might write. "Today I learned about photosynthesis."

Next, **circle a face** to show how you felt about it:

- ☺ (fun),
- 😐 (okay), or
- 😟 (confusing).

This simple **reflection (Bloom's *memory*)** helps your brain, and an **AI tutor** knows what stuck with you.

- It's **visual and tactile** – You're **writing** and **interpreting** your feelings.
- Share your sentence with the class **(auditory)** if you like!
- **Reflecting** even a little bit helps deepen understanding
    - You're telling your brain, "Hey, this was important."

Good job looking back (Reflecting) on your learning! 👏

****

GRADES 6–8 (UNDERSTANDING | WRITING & AUDITORY):

Use the **3-2-1 reflection** method **(an AI or teacher might prompt this):**

- **3 things** you **remember** or learned today.
- **2 questions** you **have or things** you're curious to explore more.
- **1 connection** or **favorite part** of the lesson.

Take a few minutes to **list them**:

- **3 Things I Learned (e.g., three facts or skills)**:
  1. ______________________________
  2. ______________________________
  3. ______________________________

- **2 Questions (anything you're wondering)**:
  1. ____________________________
  2. ____________________________

- **1 Favorite/Connection ((maybe "This reminds me of …")**:

1. ____________________________

This structure guides you through:
- R*emembering* facts,
- *Comprehending* (U*nderstanding*) what puzzles you, and
- R*eflecting* on personal relevance

(Bloom's levels from Knowledge (recall) to Comprehension (understand) to Evaluation (reflect).

It's also a **writing exercise (tactile)**, and
- You can **discuss your 3-2-1 with a partner** after **(auditory)**
- Often, their questions or favorites will spark a great conversation.

Reflection doesn't end with writing. ➰
- Talking about it can further solidify knowledge.
- AI loves this technique too: it learns from what you listen to personalize future lessons.
- You're training your brain (and indirectly an AI) by reflecting!

****

GRADES 9–10 (EVALUATING | WRITING & METACOGNITIVE):

Tool: davidsondavie.edu

Rate your understanding of today's material on a scale of 1–5 and explain why:
- **My comprehension (understanding)**:
- Scale of 1 to 5: __________/5
- **Reason:** "______________________________________."

For Example:
"3/5 – I grasp main ideas but need more practice on details," or
"5/5 – I could teach it to someone else!"

Be honest: if you put 2 or 3, that's perfectly fine – it shows you know where you need help.

Bloom's Analysis and Evaluation Skills:

- This self-assessment is a **higher-order** skill (Bloom's **analysis** and ***evaluation*** your learning).
- It's written **(tactile)** and deeply **reflective**.

Further Reflection and Refinement Skill:

Tool: davidsondavie.edu

- **Now, think about what would boost you to 5/5.**
- **Write one action you (or an AI tutor) could do to help:**

"I will ______________________________." ***(e.g., "Review key terms," or "Ask AI to quiz me on this topic.")***

Skills Mastered with AI:

- You are essentially engaging in a **reflective dialogue** with yourself (or an AI) about learning.
- **Share** your rating (anonymously) on the board if you'd like – it helps the teacher and AI understand how the class feels overall.
- **Remember,** this is **NOT for a Grade;** it's for **Your Growth**.
- **Reflection** helps turn **experience** into **learning.**– and you're doing it!

All Grades Bonus – "Ask Me" AI Mirror:

- Pretend you have an AI that asks: *"What was the most interesting thing you learned, and why?"*
- Take a moment to **answer that question in a short paragraph or doodle**.
- You can write: "The most interesting thing I learned was _____, because ______." Or
- Draw a quick sketch of that idea.
- By articulating this, you're teaching *yourself* why it stood out, which makes it stick in your memory.
- This is a mini-dialogue with an imaginary AI (or just yourself in the mirror).
- It's both visual (if you draw/write) and verbal (if you read it out loud).

****

Note: Research shows that reflecting in this way **deepens your comprehension (understanding).**

Tool: davidsondavie.edu.

- So next time you learn something cool, **remember** it better (and longer) by **reflection**!

STRATEGY TEN

CQI-GOAL SETTING WITH AI SUPPORT (PHASE IV: REFLECT)

AI can help you set and achieve learning goals by providing suggestions and reminders.

ACTIVITY:

****

GRADES 4–6 (APPLYING | WRITING & DRAWING):

- *(Bloom's Application/Planning skills) (Visual, Auditory, & Tactile)*
- Think of a **learning goal** for yourself. Maybe:
- "I want to read three books this month," or
- I want to get better at multiplication.

"**Write your goal** in one sentence:
"My goal is to: "____."

Now, list one thing you will do to reach it:
"To reach it, I will _______________________________________."

For example:
"My goal is to **learn 20 new vocabulary words.**
--To reach it, I will **practice five words each week.**"

Writing it down makes it real **(a tactile action).**

- Next, **draw a tiny trophy** 🏆 next to your goal
- That's you achieving it! T
- This **visual** adds motivation.
- Sharing time: **Share your goal with the class (voluntarily) (auditory).**
- Saying it out loud can make you feel committed.
- AI assistant asks you to state goals and steps clearly
- Keep your goal sheet; maybe an AI or your teacher will check in on your progress later. You got this!

****

GRADES 6–8 (ANALYZING | WRITING & PLANNING):

Let's create a mini–Goal Planner.
Identify an academic goal (e.g., "Improve my math grade to a B by next term").

Fill in the Planner:

- Goal: _____________________________________ (be specific and realistic).

- Steps to achieve it:
  1. ______________________________
  2. ______________________________
  3. ______________________________

- One resource or AI tool to help:
  ______________________________________.

For Example: Goal: "Finish a 5-page research paper on time."
STEPS:
1) Pick a topic by Monday,
2) Write one page each day,
3) Use spell-check/AI to edit.

Resource: "Grammarly (an AI writing checker) to polish my draft."

By detailing steps, you used Bloom's *analysis* (breaking the task down).

Listing a tool shows you're thinking of support – Maybe an AI reminder app or a tutor bot.

This plan is visual and logical.

Review your plan – Does the timeline make sense?

(If not, adjust it now – that's planning in action!)

--You can even **pair up and share plans** for feedback (auditory).
--An AI coach would love this structured approach.

GOAL-SETTING + AI HELP = A WINNING COMBO FOR SUCCESS! ✅

****

GRADES 9–10 (CREATING & EVALUATING | WRITING & METACOGNITIVE):

Formulate a **SMART goal** (Specific, Measurable, Achievable, Relevant, Time-bound) related to your studies **and** a question you'd ask an AI to help achieve it.

For Example:

- Goal: "Increase my chemistry test score by 15% in the next 2 months."
- AI Help Question: "What are effective study techniques for chemistry, and can you remind me to study 3 times a week?"

Now you try:

- Goal:
- *"I will… (include a number or deadline)__."*

- AI Help Question:
- *"What/How can AI…?"* (ask for a tip, schedule, or resource).

Maybe your goal is:

- "Read five novels by year-end."
- And, you ask AI, "Give me a reading schedule and ask me questions after each book."

By setting the goal, you exercised Bloom's *creative* (formulating a new plan) and *evaluation* (ensuring it's realistic) thinking.

- By thinking of an AI assistant, you show higher-order planning – you know how to leverage tools.

This is a written and mental exercise.

- Share your goal + AI plan in small groups (auditory);
- Others might want to borrow your clever ideas!
- Tracking goals with AI can keep you accountable
- Imagine an app cheering you on when you hit milestones.
- With (or without AI), writing goals and steps like this put you on the path to achievement.

Go conquer those goals! 🎯 🚀

ALL GRADES BONUS – DREAM BIG DRAWING (VISUAL & CREATIVE):

- **Draw yourself in the future achieving your goal,** with a friendly AI helper by your side. For instance, if your goal is to become a coder,
- **Sketch yourself at a computer with a little robot buddy handing you code tips.** Or if you want to master a test,
- **Draw yourself holding a test paper with an AI character** giving a thumbs-up.
- **Be creative – add speech bubbles or captions** if you want (maybe your AI says, *"You can do it!").*
- **Color and draw** (both visual and tactile) -- Make your goal feel more real and enjoyable.
- **Display your drawing** on the classroom wall as a reminder of what you're working toward.
- **Every time you see it,** you'll remember your goal and the fact that you have tools (and people and maybe AI) to help you.
- **Dream big,** work hard, and one day that drawing could/will become a reality! 🌟🎨.

YOU CAN DO IT! KEEP BELIEVING, ACHIEVING,

& SUCCEEDING!

CONGRATULATIONS, STUDENTS (& TEACHERS)!

You've completed the selected pages of my AI TOOLKIT WORKBOOK for STUDENTS

You've planned, learned, practiced, reflected, and set goals (under the guidance of your teacher) – all with a bit of AI magic sprinkled in.

Each activity was designed to engage your eyes, ears, hands, and mind **(VAKT)**, and to challenge different thinking skills from **Bloom's Taxonomy** (remembering facts, understanding concepts, applying skills, analyzing and evaluating information, and creating new ideas).

By completing these activities, you aren't just learning about your class subjects – you're also seeing how **AI can be a helpful partner** in your education, and how **you** can think in new ways utilizing AI and other technologies—and more importantly, how to THINK and APPLY new information with (or without) technology.

--Keep this workbook as proof of how much you've grown. And remember: the future classroom is:

YOU + AI CREATIVITY + CURIOSITY ALL WORKING TOGETHER.

--Stay curious, keep reflecting, and never hesitate to ask questions to your teachers, to your peers, or even to an AI helper.

HAPPY LEARNING & SUCCEEDING!

Note From "Aunt Hattie" - It is advisable to use modern technological tools while also keeping your mind active and remembering essential facts and data, in case of power outages or technology malfunctions. Yes, that can happen, and we need to know specific essential phone numbers (especially those of immediate family members), directions to important places (such as parents' or grandparents' homes, the hospital), etc., without relying on Alexis, our cell phones' prompts, or GPS. A TESTIMONY & REFLECTION: NO ELECTRICITY FOR 24 HOURS

AN EXTRA TREAT: CLICKABLE WALK-THROUGHS OF TWO POPULAR AI TOOLS
(MagicSchool.ai & Curipod.com)

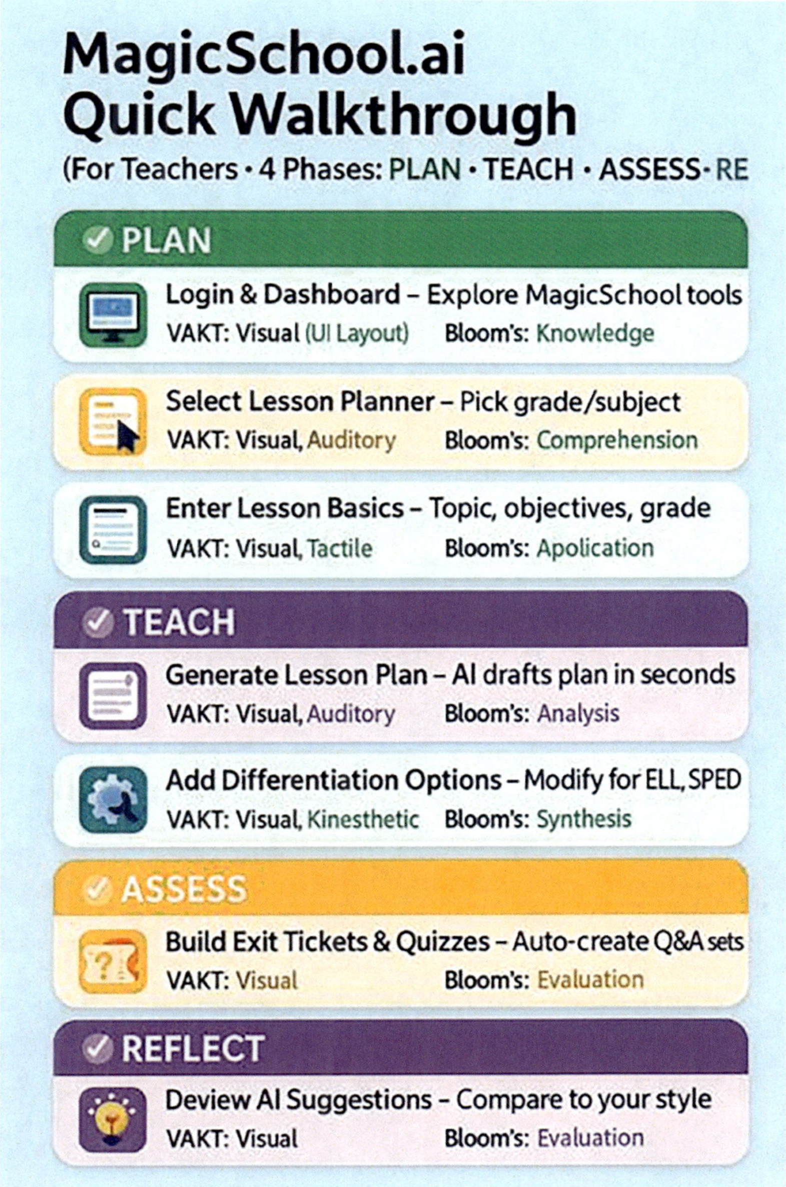

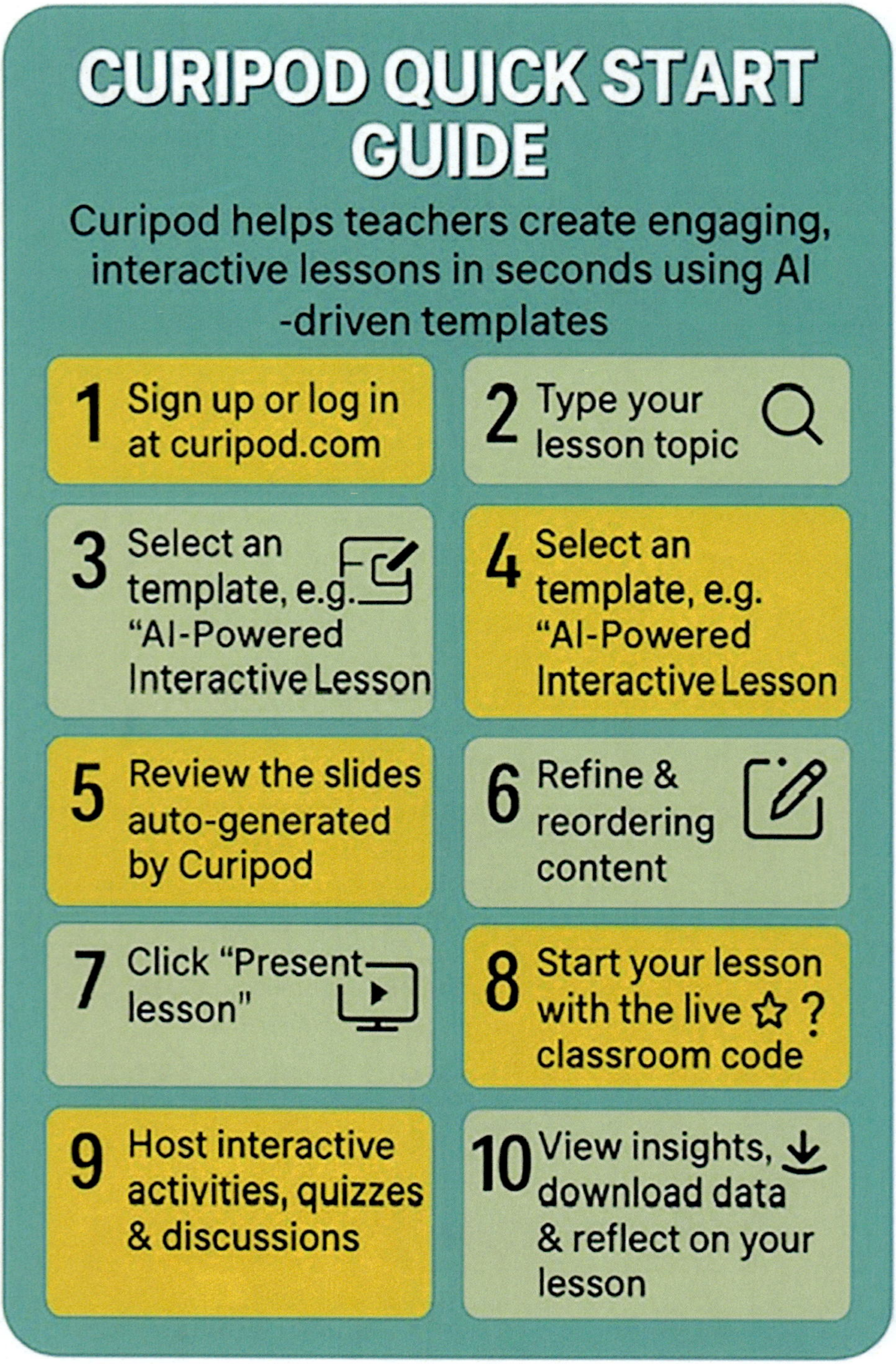
CURIPOD QUICK START GUIDE
Curipod helps teachers create engaging, interactive lessons in seconds using AI -driven templates
1 Sign up or log in at curipod.com
2 Type your lesson topic
3 Select an template, e.g. "AI-Powered Interactive Lesson
4 Select an template, e.g. "AI-Powered Interactive Lesson
5 Review the slides auto-generated by Curipod
6 Refine & reordering content
7 Click "Present lesson"
8 Start your lesson with the live ☆ ? classroom code
9 Host interactive activities, quizzes & discussions
10 View insights, download data & reflect on your lesson

MORE ENDORSEMENTS FOR DR. WASHINGTON'S AI TOOLKIT FOR TEACHERS:

"This book is a powerful blend of my mom's decades of teaching experience and her passion for staying ahead of the curve with the latest technology. She's always used innovative tools to help students reach their full potential, including my sister and me. Thanks to her approach, we became a doctor and a lawyer. I've seen the impact of her methods firsthand, and now other educators can too."
Cheryl Y. Washington, Esq.
President/CEO
East Baltimore Development Inc. (EBDI)

"Learning to leverage AI can empower teachers to navigate the modern, evolving landscape of education. By utilizing AI-driven tools, educators can automate grading, analyze student performance data to identify learning gaps, and tailor instruction to meet individual needs. AI serves as a powerful ally in preparing teachers to meet the demands of a dynamic, tech-integrated future."
Wayne Saunders
Governance, Risk & Compliance Engineer
Bethesda/Microsoft Gaming

"My mom's new book on using AI in the classroom is a timely and empowering guide for educators navigating today's ever-changing landscape. As a physician, I've seen how AI is reshaping the medical field—from diagnostics to treatment—and it's exciting to see her bring that same forward-thinking approach to education.

She's spent her life dedicated to children and preparing future teachers, and this book reflects her unwavering commitment to both. I'm so proud to see her leading the way, helping educators embrace new technology while staying grounded in what matters most: inspiring and equipping the next generation.
Charrell W. Thomas, MD"

"As a beloved and devoted school bus driver for special needs students for over 30 years, I've seen how overwhelmed teachers can get trying to meet every child's unique learning needs, especially emotional student learners (ESL).Dr. Washington's The AI Toolkit for Teachers Made Easy is the lifeline they've been waiting for. It offers simple, smart AI tools to help teachers personalize learning and save precious time without burning out. I highly recommend this book by Dr. Washington, the renowned teacher's teacher, also known as "Aunt Hattie."
Jessie Handley
A Grateful Veteran School Bus Driver for Special Needs Students."

"The AI Toolkit for Teachers Made Easy is more than a guide - it's a bridge to educational equity and innovation. As a leader of a nonprofit organization committed to empowering women and communities through connection and learning, I see this book as an essential tool for educators ready to inspire and lead in today's AI-driven classrooms and the world."
Peggy Morris
President/Founder, Sisters 4 Sisters Network, Inc.

---

*"Congratulations, Dr. Washington, on your new timely book publication, **THE AI TOOLKIT FOR TEACHERS.***
I am sure that your book will not only assist educators in the classroom, but also guidance counselors, school administrators, and mentors alike. This AI book will also be a catalyst for all students to realize their full potential while studying and competing on the global stage to become a great success, while facing educational challenges with more ease with AI assistance!"
-- Reba N. Burruss-Barnes, Publicist,
REBAssociates, RESOURCE NETWORK INT

"Dr. Washington turns AI from an intimidating buzzword into an empowering classroom ally. This book is a must-have guide for every educator determined to reach all learners by working smarter—with compassion, creativity, and courage."
-- Valerie Etienne-Leveille,
Director of Professional Learning Certification Program
St. Johns County School District, Jacksonville, FL

"The AI Toolkit for Teachers Made Easy is a timely and practical resource that makes AI accessible for all educators. Written by the highly respected Dr. Hattie Washington, this guide offers clear strategies and ready-to-use tools to help teachers personalize learning, save time, and reignite their passion for teaching. Whether you're tech-savvy or not, this empowering book is a must-have for today's classrooms."
-- Kelly King Toaston, PHR, MPA
Vice President and Chief Human Resources Officer,
Flagler College, Jacksonville, FL

---

*"It is with esteem and pleasure to share my endorsement for Dr. Hattie Washington's new book called '**AI Toolkit for Teachers Made Easy**'! This is a timely teacher's tool from a true educator, scholar, and living ancestor who works tirelessly every day to add value for the betterment of future generations. Dr. Hattie is equipping children, especially those from disadvantaged backgrounds, with state-of-the-art tools, techniques, and skills to compete for today's most in-demand jobs, which are essential. I see this AI tool taking our children to the next level of greatness, and I'm eager to promote this much-needed book to the world! Truly,"*
-- Sharon Parker,
Leading Within Strategist!"

"Dear Dr. Hattie,
I am deeply honored to endorse Dr. Hattie N. Washington's groundbreaking book, The AI Toolkit for Teachers Made Easy. As a CEO in real estate and construction, as well as a wellness coach, I recognize the importance of simplifying complex tools to empower individuals in all areas of their lives. Dr. Washington's book does exactly that for educators, offering practical, actionable strategies that make artificial intelligence both accessible and transformative in the classroom. In today's fast-paced world, where time is a precious commodity, Dr. Washington provides teachers with the tools they need to save time, personalize learning, and foster stronger connections with their students, without requiring technical expertise.

I am incredibly grateful for Dr. Washington's dedication to equipping educators with the confidence and resources they need to thrive in a world driven by technology. This book is more than just a guide; it's a powerful resource that will help teachers navigate the future of education with ease and purpose. In an era where AI is becoming a critical component of every industry, Dr. Washington ensures that educators are not left behind but are instead empowered to utilize these tools effectively. In my international consultancy and inspirational speaking, I am integrating AI on a global stage, which is undoubtedly significantly impactful in supporting individuals and organizations in embracing technology and achieving transformative results. Dr. Washington's work is a game-changer for education, and I look forward to seeing the profound impact it will have on teachers and students alike."
--Sabine Vicioso David,
Founder & CEO, Lunovi Real Estate in Punta Cana,
Dominican Republic Real Estate

"I am honored to endorse Dr. Washington's transformative book, The AI Toolkit for Teachers Made Easy. As an AI business and automation solutions expert based in Tucson, Arizona, I recognize the profound impact that effective technology integration can have on education. Dr. Washington's book provides educators with a clear roadmap to harness the power of artificial intelligence, enabling even those without a technical background to navigate this essential landscape with confidence. In a world where every moment counts, Dr. Washington equips teachers with straightforward, actionable strategies that not only save time but also personalize learning experiences for every student. This toolkit is not just a resource; it's a lifeline for educators striving to create meaningful connections in the classroom amidst the complexities of modern teaching.

Dr. Washington's dedication to empowering educators is evident throughout the book. Her insights are not only timely but also vital for those who wish to thrive in an evolving educational environment. I wholeheartedly believe that The AI Toolkit for Teachers Made Easy will be a transformative asset for teachers, enabling them to embrace AI without feeling overwhelmed. I am excited to see how Dr. Washington's innovative approaches will inspire educators and students alike, paving the way for a brighter future in education."
Lamar Myers AI Business and Automation Solutionist Tucson, Arizona

"From someone in the IT field with experience with AI, this is a great roadmap for teachers. It's easy to implement and understand, and very empowering."
— **Ivan M Johnson**, CEO Novel Applications of Vital Information, Inc (NAOVI)

"Dr. Hattie Washington shows us how to teach smarter, not harder—with compassion, creativity, and courage—empowering every student to discover their strengths, sparking curiosity in the classroom, and inspiring educators to lead with purpose."
— **Shurvone Wright**, La'Boss Entrepreneur Academy of Creative Solutions

"This book is much needed in the K-12 classroom as AI has become the trend in our educational system. As an Autism paraprofessional in the SPED Department, her book will help build the life skills that are needed."
— **Ms. Aileen J. Hill**, Teacher | Best-Selling Author | International Speaker

"Excellent! Dr. Washington, why did you wait until my retirement to write this book? This AI roadmap is invaluable for non-tech-savvy administrators like myself. It's also an excellent tool for all educators. As a former Title I school administrator, I highly recommend incorporating this book into your educational resources to help teachers personalize learning for all students' diverse needs. This book is a game-changer!"
— **Dr. De'Rain F. Irvin,** Retired Administrator, CEO 3N1MINISTRIES, LLC & Executive Pastor - Hope Aglow Empowerment Church

"An innovative toolkit that empowers educators of all backgrounds to harness the power of AI, even if they're not tech enthusiasts." "Dr. Hattie Washington reveals a transformative approach to education emphasizing how to teach with creativity, bravery, and empathy for maximum effectiveness."
— **Barry S. Ricoma,** Quality Assurance Specialist

"The AI Toolkit for Teachers Made Easy" by Dr. Hattie N. Washington is a groundbreaking, must-have guide that every educator, parent, and school library should own. Dr. Washington masterfully demystifies AI with clarity and heart, offering 10 powerful strategies that even the least tech-savvy teacher can implement to transform classrooms and ignite student success. As an advocate for leadership and empowerment, I am deeply inspired by this book's originality and practical brilliance. It is not only a timely resource, but it is a movement toward educational equity and excellence. Get this book. Your students, your children, and our future will thank you."
— **Dr. Madeline Ann Lewis**, President/CEO, Executive Women's Success Institute

"Dr. Hattie Washington's New Book: THE AI TOOLKIT FOR TEACHERS MADE EASY will be "A timely toolkit that makes AI accessible to all educators—especially those who've never felt tech-savvy." I understand the importance and use of AI in the world today. It's a must know."
— **Dee Sands**,
Primerica Representative

"Thank you, Dr. Hattie, for this powerful tool—an essential guide that not only equips teachers to help students succeed but also helps parents get ahead and better support their children and teachers alike. Every parent should read it to strengthen the partnership between home and classroom."
— **Mark Wiggins**, "The SpeakerMan" Editor-in-Chief, "OFF the BENCH" Magazine & Podcast

"Dr. Hattie N. Washington's book, The AI Toolkit for Teachers Made Easy, offers clear, actionable strategies that empower educators to use AI effectively—regardless of their tech background. She communicates in indisputable terms, making AI practical, ethical, and accessible. This resource is especially vital for school boards focused on equity and closing the digital divide in underserved communities."
— **Jacqueline (Pastor Lulu) Brown**, Founder & CEO, Revolution Ascension LLC, Retired Global Technology Executive | Licensed Minister Transformation Speaker | Consultant & Coach

"Dr. Washington makes the impossible feel possible. This toolkit doesn't overwhelm—it uplifts, meeting teachers where they are and guiding them into the future with grace."
—**James LeGrand**, Author of Evolve! and 10 Things I Told My Kids About Trayvon Martin and George Zimmerman; Co-Author of Ignorance is Contagious and The Reluctant Witness

"This is the AI roadmap every teacher needs. Practical, empowering, and full of heart."
— **Nadia Francois**, CEO of Heiress International

"Dr Hattie Washington had designed an AI Toolkit that will help K-12 teachers as well as other community leaders to teach smarter and not harder. A treasure chest of gems for the NOW teachers."
— **Dr. Berlinda A.** Hart Love

"When you combine passion with knowledge and add in technology -- the outcome is pure genius, and that's exactly what this book is! This book positions educators to create pedagogy that will take students to the next level and beyond. Kudos Dr. Hattie."
— **Dr. Renata A. Walton,** THE WALTON WAY

Dr. Hattie Washington demonstrates the smart approach to learning AI!
— **Lawrence Stokes**, Best-Selling Author | International Speaker

Placing this book in the hands of every educator across the US will reduce the disparity gap in education and give every student the opportunity to succeed. Thank you, Dr. Washington.
— **Elaine Y. Scales**, EDS Integrative Services, LLC - Mindfulness & Meditation Teacher | Transformational Coach | Sound Healer & Teacher | Mental Health First Aid Instructor

"This priceless toolkit gives open doors of creativity, an OPEN invitation to NEXT-LEVEL LEARNING!"
— Rev. J.A.Mack Sr., USMC/DAV

"AI is a powerful educational tool when it is partnered with the "human touch". Dr Hattie Washington successfully marries the two."
— Teresa A. Palmer, MSN, RN, NPC

"If we want our students to thrive in a future driven by AI, we must first equip the ones guiding them—our teachers. This toolkit does exactly that, with clarity, care, and real-world application. I wholeheartedly recommend this timely and transformative work to every educator committed to shaping the next generation of thinkers, leaders, and innovators."
— Jose Otero, Father | Grandfather | Software Development Professional 30+ years

"Dr. Hattie Washington, it was such a pleasure meeting you in the Dominican Republic. Your book, "The AI Toolkit for Teachers Made Easy" is more than just a resource for educators; it is a lifeline for future-shaping classrooms. This book bridges the gap between innovation and accessibility, offering practical, clear, and empowering strategies for teachers—especially those who may not see themselves as 'tech-savvy.' It removes the intimidation from artificial intelligence and replaces it with confidence, creativity, and connection."
— Dr. Reginald Siles,
Spirit of A Dove (SOAD) & 20. Mrs. Sharon Siles, RN

"As a single mother to a teenage daughter who's about to start high school, I know how important it is for education to meet students where they are. My daughter is incredibly creative, hands-on, and full of ideas—but if she's not challenged, she checks out. She needs learning that feels relevant, engaging, and forward-thinking. The AI Toolkit for Teachers Made Easy is exactly what teachers need to spark curiosity and keep students like my daughter inspired. It shows educators how to use AI to personalize learning, make lessons more interactive, and bring out the brilliance in every student—especially those who think outside the box. This book doesn't just prepare teachers—it helps prepare our kids for a world that demands creativity, innovation, and adaptability. I'm grateful for this resource and the future it's helping to shape."
— Ms. Trina A. Holmes-Otero,
Single Mother of a Creative Teen, East Coast USA

"As a father of young children, and a professional in the fast-moving tourism industry, I understand how quickly the world is changing—and how critical it is for education to evolve alongside it. We're not just raising children; we're preparing them for careers that don't even exist yet. Our children can't afford for schools to fall behind. The AI Toolkit for Teachers Made Easy is a game-changer—giving teachers the confidence to use AI now, not later.
— Jose A. Flores, Father | Tourism Industry Professional - Dominican Republic

"As a single mother to a baby who hasn't even started school yet, I'm already thinking about the future my child will face. The world is changing fast, and education needs to keep up. The AI Toolkit for Teachers Made Easy gives me hope. It shows how teachers can embrace technology without fear—so our kids won't be left behind. I may not be in the classroom, but I know this book is exactly what our educators need to prepare the next generation—starting now."
— **Ms. Judith Stephanie De León De Jesús**, Single Mother, Dominican Republic

"Mucho gusto, Dr. Washington! It is an honor to endorse your book, "The AI Toolkit for Teachers Made Easy." As a university student native to and living in the Dominican Republic, I've seen how quickly technology is changing the world—and how important it is for our teachers to keep up. But I've also seen how hard it can be, especially when many educators feel overwhelmed or unsure about using new tools like artificial intelligence. Your book is exactly what our schools need. It speaks to teachers in a way that's simple, encouraging, and real. This book doesn't just teach about AI—it gives practical strategies that can make learning more personal and powerful for students like me, no matter where we're from."
— **Ms. Scarlet Noelia Ortiz Suazo**, University Student, Dominican Republic

Dr. Hattie N. Washington

Dr. Hattie N. Washington is an internationally recognized educator, award-winning author, and pioneer in technology-driven learning. Over 30 years ago, she hosted the TV show, Computers for Kids, bringing tech literacy to underserved families across the DMV (DC, Maryland, and Virginia) before digital learning became mainstream. She launched after-school computer labs in empowerment-zoned churches and installed computer centers in her three foster boys' homes, aligning instruction with state standards, promoting critical thinking, and accommodating diverse learning styles.

A proud mother of two accomplished daughters—a physician and an attorney, she also reared scores of foster boys with special needs at a nonprofit she founded, **Aunt Hattie's Place**, where she's affectionately called **"Aunt Hattie."** She inspired her first foster son to pursue a career in IT, who is now an IT specialist at Microsoft. Her career includes 25+ years as a model teacher, Assistant Superintendent of Baltimore Schools, and State Evaluator of Superintendents of Title I Schools, tenured Professor—training teachers, and first female Vice President at Coppin State University (CSU).

Dr. Washington holds a BS Degree in Elementary Education & Special Education; a Master's in Counseling Psychology; and a Ph.D. in Curriculum Instruction. Further Study: Special Education & Multiculturalism, Glasgow University, Scotland, UK; Executive Leadership & Management, Harvard; & Innovation Techniques in Higher Education, London Roundtable. She is a member of AAAI (Association for the Advancement of Artificial Intelligence) and is completing training as an AI Consultant.

She is the author of four award-winning/best-selling books: AUNT HATTIE'S COOKBOOK: Southern Comfort Food Favorites; DRIVEN TO SUCCEED: An Inspirational Memoir of Lessons Learned Through Faith, Family, and Favor; and DETERMINADA A TRIUNFAR (DRIVEN translated into Spanish)--*Partial proceeds from sales support* ***Aunt Hattie's Scholarship Fund (AHSF)*** *to help with college expense for former foster youth aged out of care.*

Her recent Amazon #1 Best-seller book, THE AI TOOLKIT FOR TEACHERS MADE EASY: 10 Simple Strategies To Save Time, Personalize Learning, And Empower Every Student To SUCCEED—Even If You're Not Tech Savvy, empowers Teachers to help all students Believe, Achieve, and SUCCEED using innovation and technology. A 2018 inductee into the Maryland Women's Hall of Fame, Dr. Washington continues to be a leading voice in empowering educators, parents, and students to succeed, now using aligned AI-powered tools.

—visit: www.aitoolkit4teachers.ai or www.driven2succeed.com.

"Train up a child in the way he should go: and when he is old, he will not depart from it."

— Proverbs 22:6

NOTE: Teachers shaping learners' habits and values matter —especially when AI tools influence learning pathways.

THE AI TOOLKIT FOR TEACHERS MADE EASY:

10 Simple Strategies to Save Time, Personalize Learning, Empower Every Student to SUCCEED--Even If You're Not Tech-Savvy

Thank you! For buying this book, giving it 5 stars, and leaving your Review on Amazon.
Enjoyed this book? Recommend it to a friend.

MORE BOOKS BY Dr. Hattie N. Washington:
visit www.aitoolkit4teachers.ai or
www.driven2succeed.com.

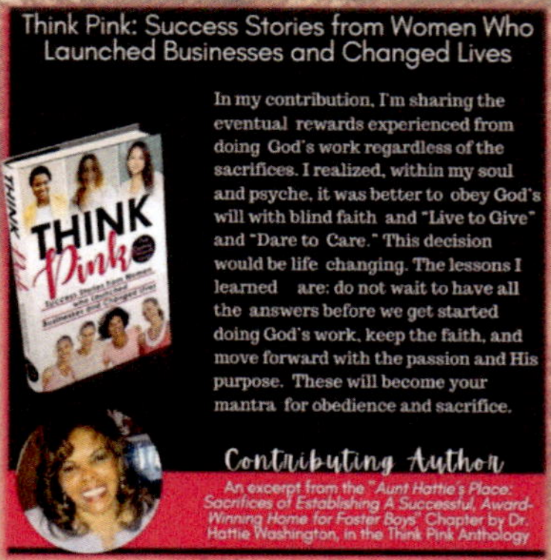

NOTE: Links to the free AI tools in the print book must be typed into your browser, while the e-book contains mostly live, clickable links. Consider getting the e-book as a companion to this book to avoid errors typing — and if a typed link doesn't work, check for typos and try again. 😉👍✅

WASHINGTON PUBLISHING™ ENTERPRISES

Jacksonville • Silver Spring • Dominican Republic

Made in United States
Orlando, FL
04 November 2026

90206043R00143